THE HYPERSTORE MANIFESTO

ON DATA, EXECUTION & STRUCTURAL TRUTH

THE HYPERMODERN THEOREM
BOOK 1

DOM JOCUBEIT

REGISTERED OFFICES

Hypermodern Limited Company, 1209 Mountain Road PL NE, STE N, Albuquerque, NM 87110, USA

For details of our global editorial offices, customer services, and more information about Hypermodern products visit us at www.hypermodern.ai.

Hypermodern also publishes its books in a variety of electronic formats and by print-on-demand. Some content that appears in standard print versions of this book may not be available in other formats.

TRADEMARKS

Hypermodern and the Hypermodern logo are trademarks or registered trademarks of Hypermodern Limited Company and/ or its affiliates in the United States and other countries and may not be used without written permission. All other trademarks are the property of their respective owners. Hypermodern Limited Company is associated with some products mentioned in this book.

LIMIT OF LIABILITY/DISCLAIMER OF WARRANTY

While the publisher and author have used their best efforts in preparing this work, they make no representations or warranties with respect to the accuracy or completeness of the contents of this work and specifically disclaim all warranties, including without limitation any implied warranties of merchantability or fitness for a particular purpose. No warranty may be created or extended by sales representatives, written sales materials, or promotional statements for this work. This work is sold with the understanding that the publisher is not engaged in rendering professional services. The advice and strategies contained herein may not be suitable for

For Lachlan

Who taught me, very early on,

that things should work the way they say they do.

Intelligence fails when truth must
be reconciled after execution.

CONTENTS

The Hypermodern Theorem 17
Preface 19
Introduction 25

PART I

1. THE COMPLEXITY CRISIS 33
The Integration Tax 34
AI Has Accelerated the Crisis 35
The Privacy and Compliance Wall 36
The Unspoken Truth 37
A New Paradigm Is Needed 38

2. WHY TRADITIONAL DATABASES CAN'T EVOLVE FAST ENOUGH 41
The Relational Plateau 42
Document Stores: Flexibility That Cuts Both Ways 42
Search Engines: A Separate World With Separate Truths 43
Vector Databases: The Newcomer With a Narrow Worldview 44
Graph Databases: Great for Relationships, Bad for Everything Else 45
Object Stores: The Last Resort Storage Layer 46
Polyglot Persistence: A Beautiful Idea That Doesn't Work in Practice 47
The Hard Limit: Databases Cannot Evolve Fast Enough 48
The Architectural Mismatch Has Become Terminal 49

3. THE NEW REQUIREMENTS OF MODERN SYSTEMS 51
Multi-Model Data as the New Default 51
AI-Native Workflows Require Deep Data Unification 53
Privacy-by-Design Is No Longer Optional 54

Post-Quantum and Advanced Cryptography Requirements 55
Multi-Tenant SaaS Has Become the Baseline 56
Provenance and Auditability Are Now Functional Requirements 57
Deterministic Backend Generation Is Becoming Necessary 58
Zero-Trust Architecture Requires Data-Level Authority 59
Unified Context Is the New Competitive Advantage 59
The Requirements Are Clear 60

4. THE DEVELOPER'S DILEMMA, REVISITED 63
The Cognitive Overload of Fragmented Architectures 64
The False Choice Between Flexibility and Safety 65
Schema Drift: The Invisible Enemy 66
The Fragility of Distributed Governance 67
The Glue Code Trap 68
The Build-vs-Scale Paradox 68
AI Workflows Create Unmanageable Developer Burden 69
Developers Are Carrying the Architecture on Their Shoulders 70
The Need for a Unifying Architecture 71

PART II

5. THE HYPERSTORE INVENTION 77
The Core Premise: Unification, Not Aggregation 78
The Hyperstore Stack Model 79
The Hyperstore Schema 81
The Hyperstore Query Model 82
Self-Governance: The Hallmark of a Hyperstore 83
The Hyperstore Performance Model 83
The Hyperstore: A New Trust Foundation 84
Why a Hyperstore Is Not "Just a Better Database" 84
The Hyperstore in Context 85
Conclusion 86

6. WHY HYPERSTORES ARE ARCHITECTURALLY POSSIBLE NOW 87
The Rise of JSON and Universal Structured Data 88
The Maturation of Strongly Typed Schema Languages 88
Cloud-Native Infrastructure Made Vertical Integration Feasible 89
The Explosion of AI and Vector Embeddings 90
Privacy, Governance & Compliance Became Core System Requirements 91
The Zero-Trust Security Movement 92
Advances in Applied Cryptography 92
Developer Expectations Have Evolved Beyond Databases 93
Distributed Systems Theory Finally Stabilized 94
The Infrastructure Stack Collapsed Enough to Build Upwards 94
The Stack Tax Became Unbearable 95
The Moment of Convergence Has Arrived 96
Conclusion 97

7. THE HYPERSTORE ARCHITECTURE 99
The Architectural Premise: One Schema, One Engine, Many Modalities 99
The Layered Architecture of a Hyperstore 100
Cross-Layer Principles 104
Emergent Behaviors Unique to Hyperstore Architecture 107
Why This Architecture Is a Break from the Past 109

8. THE HYPERSTORE SECURITY MODEL 111
The Failure of Perimeter Security 112
Declarative Security Guarantees 112
Field-Level Encryption 113
Homomorphic Computation 113
Differential Privacy 114
Post-Quantum Cryptography 114
Provenance and Cryptographic Integrity 115
Selective Disclosure and Zero-Knowledge Extensions 115
Security Across the Execution Engine 116
Multi-Tenancy and Isolation 116

Security as a Category-Level Guarantee 117
Summary 117

9. THE HYPERSTORE EXECUTION ENGINE 119
Unified Query Graphs 119
Deterministic Planning and Execution 120
Query Planning Across Modalities 121
Index-Aware Execution 121
Write-Ahead Logging and State Transitions 122
Transactions and ACID Semantics 122
Caching and Materialization 123
Distributed Execution Considerations 123
Why Hyperstores Are More Optimizable 124
Execution as a Category-Level Property 124
Summary 124

10. FLUENCY: THE HYPERSTORE API CONTRACT 127
The Hyperstore as a Programmable Endpoint 128
Schema-Derived Contracts 128
Type Systems as Guarantees 129
Deterministic Client Generation 129
Binary Protocols and Efficiency 130
Errors as Part of the Contract 130
Streaming, Subscriptions, and Reactivity 131
Versioning and Evolution 131
Fluency as a Category-Level Property 132
Summary 132

PART III

11. VEKTAGRAF: THE FIRST HYPERSTORE 137
The Design Goals of Vektagraf 138
Core Capabilities, Unified 138
The Vektagraf Schema as a Hyperstore Program 139
Execution: Making Guarantees Real 139
Security and Privacy in Practice 140
Multi-Tenancy as an Architectural Primitive 140
Fluency Realized 140
Vektagraf as a Reference Implementation 141
Summary 141

12. INSIDE VEKTAGRAF 143
Layered, but Not Fragmented 143
The Unified Storage Layer 144
Indexing as a Semantic Extension 145
The Write-Ahead Log (WAL) 145
The Execution Engine 146
Schema Compilation and System Generation 146
Multi-Tenancy in the Architecture 147
Cryptography as Infrastructure 147
Why This Architecture Is Necessary 148
Summary 148

13. THE VEKTAGRAF SCHEMA SYSTEM 149
Schema Kinds: Extensions and Applications 149
Imports and Composition 150
Types: Primitive and Declared 150
Models as Semantic Objects 151
Relationships Without Foreign Keys 152
Inheritance and Extension 152
Templates and Field Inclusion 152
Multi-Tenancy and Provenance 153
Indexing as Declaration 153
Validation, Automation, and Lifecycle 153
Internationalization and Presentation 154
The Schema Compiler 154
Why This Matters 154
Summary 154

14. AUTOMATIONS, VALIDATION, AND DOMAIN LOGIC 157
Validations as Invariants 157
Deterministic Validation Execution 158
Error Semantics and Messages 158
Automations as Declarative Behavior 159
Safe and Bounded Execution 159
Lifecycle Hooks and State Machines 160
Domain Logic Without Application Coupling 160
Testing and Verification 160
Why This Belongs in the Hyperstore 161
Summary 161

15. PRIVACY BY DESIGN 163
Privacy by Design vs Privacy by Policy 163
Data Minimization in the Schema 164
Field-Level Confidentiality 164
Encryption as Default, Not Exception 165
Selective Disclosure 165
Differential Privacy and Aggregation 165
Tenant Isolation as Privacy Boundary 166
Provenance and Accountability 166
Right to Erasure and Lifecycle Control 167
Privacy as a Hyperstore Property 167
Summary 167

16. MULTI-MODEL QUERYING IN VEKTAGRAF 169
Queries as Unified Intent 169
Structured Queries 170
Graph Traversal as Native Querying 170
Vector Similarity Search 171
Full-Text and Pattern Matching 171
Querying Encrypted and Protected Fields 172
Composability Across Modalities 172
Deterministic Planning and Optimization 172
Transactions and Consistency 173
Why This Matters 173
Summary 173

17. BENCHMARKS AND PERFORMANCE CHARACTERISTICS 175
Why Conventional Benchmarks Fall Short 176
Measuring Capability Density 176
Baseline Performance Characteristics 177
Multi-Model Query Performance 177
Impact of Governance and Security 178
Indexing and Optimization Effects 178
Scaling Characteristics 179
Interpreting Benchmark Results 179
Toward Standard Hyperstore Benchmarks 180
Summary 180

PART IV

18. THE FIRST HYPERSTORE APPLICATIONS 185
Recommendation Engines Without Data Pipelines 186
Healthcare and Privacy-Critical Systems 186
Multi-Tenant SaaS Platforms 187
Knowledge Graphs Without a Separate System 187
Enterprise Security and Governance Platforms 188
Why These Applications Emerge First 188
The Pattern: Fewer Systems, Stronger Guarantees 189
Summary 189

19. WHAT HYPERSTORES ENABLE THAT DATABASES CANNOT 191
Semantic-Native Systems 191
Privacy-Enforced Computation 192
Continuous Schema Governance 193
Zero-Integration Architectures 193
Unified Concurrency Models 194
Elimination of Accidental Complexity 194
Why Databases Cannot Catch Up 195
The Strategic Implication 195
Summary 196

20. ECONOMIC AND ENGINEERING IMPACT 197
Cost Compression via Unification 197
Developer Productivity and Cognitive Load 198
Faster Time to Correctness 199
Reduced Security Blast Radius 199
Operational Simplicity 200
Maintainability and Longevity 200
Organizational Impact 201
The Compounding Effect 201
Summary 201

21. THE FUTURE OF HYPERSTORES 203
Near-Term Predictions 203
The Decline of the Polyglot Default 204
Research Avenues 204
Toward Standardization 205
Other Hyperstore Implementations 205
The Shift in Engineering Practice 206

Organizational and Educational Impact 206
A World with Mainstream Hyperstores 207
Summary 207

22. OBJECTIONS, MISCONCEPTIONS, AND REBUTTALS 209
"This Is Just Another Multi-Model Database" 209
"This Belongs in Application Code" 210
"Schemas Will Become Too Complex" 211
"Performance Will Suffer" 211
"This Is Too Opinionated" 212
"It Won't Fit Legacy Systems" 212
"This Is Reinventing the Database" 213
"Only Large Teams Will Benefit" 213
"This Sounds Too Good to Be True" 213
Summary 214

Epilogue 215

About the Author 219
Glossary 221

THE HYPERMODERN THEOREM

This book addresses the first domain of the Hypermodern Theorem: data.

The theorem begins with a constraint: no system can behave deterministically if the data it relies on is fragmented, inferred, or reconciled after the fact. Intelligence, human or artificial, cannot be made reliable when truth is negotiated downstream.

The Hyperstore Manifesto isolates this failure mode and treats it as foundational. Before questions of authority, legitimacy, or coordination can be meaningfully resolved, systems must agree on what exists, what has occurred, and what state they are in; without interpretation.

This book therefore makes a narrow claim: unified, authoritative data is a prerequisite for any system that intends to execute rather than merely describe.

The arguments here are not abstract. They are tested through Vektagraf, a unified data engine designed to eliminate reconciliation, ambiguity, and narrative truth in favor of executable state. Vektagraf is not offered as proof of correctness, but as a falsifiable implementation of the claims advanced in this book.

If data cannot be made authoritative without trust, discretion, or governance, then the Hypermodern Theorem fails at its first step.

If it can, the remaining domains, authority and coordination, become mechanically addressable.

This book stands alone. But it is not isolated.

It is the base layer.

PREFACE

This book was not planned.

It emerged gradually, out of frustration with systems that were technically sophisticated yet structurally brittle, systems that worked until they didn't. Systems where correctness depended on discipline, privacy depended on policy, and meaning depended on everyone remembering the same unwritten rules.

Over time, a pattern became impossible to ignore: most of the complexity in modern software is not inherent to the problem domain. It is the byproduct of dividing responsibility across too many layers, tools, and abstractions that do not share guarantees.

The Hyperstore concept arose from asking an unfashionable question:

What if the data system took responsibility for more, not less?

Not more features, but more guarantees.

THIS BOOK IS A MANIFESTO

Not in the sense of ideology or persuasion, but in the original sense: a declaration that a category no longer fits its purpose, and that a new one must exist.

For decades, we have treated data systems as passive infrastructure, repositories that store facts and retrieve them efficiently, while meaning, behavior, trust, and correctness are imposed elsewhere. That division once made sense. Today, it is the primary source of fragility in modern software.

This book makes a categorical claim:

No system can behave deterministically if the data it relies on is fragmented, reconciled after execution, or governed outside the data layer itself.

That claim situates this book as **Book 1** of a larger construct: **The Hypermodern Theorem**.

POSITION WITHIN THE HYPERMODERN THEOREM

This book occupies a specific position within the Hypermodern Theorem.

The theorem itself is defined elsewhere in this volume. It is not re-derived here.

This book addresses the **first and foundational domain** of the theorem: **data**.

Its concern is narrow and structural. Before authority can be enforced, before coordination can scale, and before intelligence can be trusted, a system must be able to determine, mechanically and without reconciliation, what exists, what has occurred, and what the current state is.

The argument advanced in the chapters that follow is that existing data architectures cannot satisfy this requirement, not because of flawed implementation, but because of structural fragmentation.

The Hyperstore is introduced as a response to that specific failure mode.

THE CLAIM OF THIS BOOK

This manifesto advances a narrow, testable claim:

Unified, authoritative, multi-model data is a prerequisite for any system that intends to execute rather than merely describe.

From that claim follows the invention of a new architectural category: the **Hyperstore**.

A Hyperstore is not a better database. It is a unified data architecture in which storage, indexing, validation, security, provenance, automation, and API generation are expressions of a single schema-driven system.

This book does not ask the reader to accept that claim on faith.

VEKTAGRAF AS FALSIFIABLE IMPLEMENTATION

Vektagraf appears in this book for one reason: **falsifiability**.

A manifesto without instantiation is speculation. A theory without construction is rhetoric.

Vektagraf is presented as a concrete implementation of the Hyperstore model, designed to test whether unified, self-governing data is architecturally possible in practice. It is not offered as proof of inevitability, market success, or optimal design.

It is offered as a system that can fail.

If unified data cannot enforce schema consistency, privacy guarantees, provenance, and multi-tenancy without external governance, then the Hyperstore concept fails.

If it can, then the first domain of the Hypermodern Theorem holds, and the remaining domains become mechanically addressable.

WHO THIS BOOK IS FOR

This book is written for readers who build and reason about complex systems:

- Database engineers and architects
- Backend and platform engineers
- AI and data infrastructure builders
- Security- and privacy-conscious system designers

It assumes familiarity with modern data architectures, but it does not assume allegiance to any particular technology stack or ideology.

HOW TO READ THIS BOOK

The book is structured in four parts:

- **Part I** motivates the problem and the limits of existing systems
- **Part II** defines the Hyperstore as a new category
- **Part III** examines Vektagraf as a concrete realization
- **Part IV** explores impact, applications, and future implications

While the argument is cumulative, readers may find value in moving non-linearly, returning to earlier chapters as later ideas crystallize.

A NOTE ON TONE

Some claims in this book are deliberately strong. I make no apologies for this.

It is not because alternatives are foolish or obsolete, but because incremental language cannot describe categorical change. The relational model was once controversial. So were transactions. So was the idea that schemas mattered at all.

If certain statements provoke disagreement, that is a feature, not a flaw. New categories deserve rigorous challenge.

AN INVITATION

The Hyperstore thesis is an invitation to think differently about where meaning lives, where guarantees belong, and what we should reasonably expect from our data systems.

You do not need to agree with every conclusion to find value here.

If this book sharpens your questions about correctness, privacy, or architectural responsibility, then it has succeeded.

Everything else follows from that.

WHAT THIS BOOK IS – AND IS NOT

This is not a tutorial.

This is not a product manual.

This is not an incremental proposal.

It is a manifesto about **architectural responsibility**.

It names a failure mode the industry has normalized.

It defines a category that resolves that failure structurally.

It grounds that category in a falsifiable implementation.

This book stands alone.

But it is not isolated.

It is **Book 1**.

INTRODUCTION

Every generation of software inherits assumptions it no longer questions.

For decades, one such assumption has been that databases are passive: they store data, retrieve it efficiently, and leave meaning, behavior, and trust to layers above. As systems grew larger and more distributed, we accepted the consequences, sprawling data pipelines, duplicated validation logic, fragile security boundaries, and an ever-widening gap between what systems *intend* to do and what they can *guarantee*.

This book begins from a quiet but radical premise: **that gap is not inevitable**.

The Hyperstore thesis argues that meaning, behavior, and guarantees belong with data itself, not scattered across applications, services, and human process. When schemas become executable, when security and privacy are structural, and when execution is unified rather than fragmented, entire classes of complexity simply disappear.

This is not a book about a product.

Vektagraf appears in these pages because ideas must eventually confront reality. It serves as a concrete proof that the Hyperstore is not a thought experiment, but a buildable system. Yet the deeper purpose of this work is to name a category, define its principles, and make its implications explicit, so that others can build, critique, and extend it.

If you are a database practitioner, this book will challenge familiar boundaries around what a data system is responsible for.

If you are a systems architect, it will offer a way to collapse complexity that has long been treated as unavoidable.

If you are a builder of AI-driven, privacy-sensitive, or multi-tenant systems, it may explain why your current stack feels heavier than it should.

New categories rarely announce themselves loudly. They emerge when accumulated workarounds become heavier than the problems they were meant to solve.

The Hyperstore is such a moment.

This book does not ask you to abandon hard-earned skepticism. It asks you to apply it rigorously, to examine whether the assumptions we have normalized still serve us, and whether a stricter, more declarative foundation might actually be the more flexible one.

If, by the end, you find yourself thinking less about tools and more about guarantees, then this book has done its job.

THE HYPERSTORE MANIFESTO

PART I

WHY THE WORLD CAN NO LONGER RELY ON THE MULTI-DATABASE STACK

Modern software is collapsing under the weight of its own infrastructure.

Applications that once required a single database and a handful of services now depend on sprawling, multi-layered stacks: a relational or document database, a cache, a search engine, a vector store, a graph engine, an object store, a message queue, a workflow engine, a schema registry, an identity service, an audit log, a key management system, and sometimes multiple analytics systems bolted on top.

Each component solves one part of the problem.

None solves the whole problem.

And the developer is left stitching it all together.

This architectural fragmentation has become the silent failure mode of the software industry. What began as specialization, using the right tool for the right job, has expanded into systemic complexity. Today's backend architectures resemble multi-organ organisms assembled from unrelated species, with overlapping responsibilities, incompatible assumptions, and inconsistent security models.

Every year, the complexity grows.

Every year, the burden shifts further onto the developer.

And no one is stepping back to ask a basic question:

Why are we building systems this way?

Complexity is what systems accumulate
when responsibility is unclear.

CHAPTER 1
THE COMPLEXITY CRISIS

When a new application is designed today, the "default" architecting process reads like a checklist of competing tools:

- A document database for flexible storage
- A search engine for text and keyword search
- A vector database for embeddings and semantic search
- A graph database for relationships
- An object store for files and binaries
- A key management store for encryption and secrets
- A schema registry for validation
- A workflow engine for automations
- A queue or pub/sub for events
- A log store for provenance and audit
- A separate analytics engine for insights
- A microservices layer to bind it all together

This is the normal path.
This is the accepted pattern.
This is, increasingly, the problem.

The number of moving parts continues to grow because each new requirement: AI, privacy, security, compliance, real-time collaboration, tends to create a **new class of database**, rather than strengthen existing ones.

So teams keep adding systems. And with each addition, the architecture becomes more brittle.

THE INTEGRATION TAX

Integrated systems aren't really integrated; they are stitched.

Every connective tissue, API wrappers, sync jobs, ETL, schema adapters, indexers, background processes, becomes a point of failure. Across dozens of tools, organizations accumulate an invisible but massive operational cost:

The stack tax.

It shows up as:

- Outages from failed syncs
- Silent data corruption from divergent schemas
- Security gaps between services
- Performance cliffs from cross-system queries
- Latency waterfalls
- Developer onboarding complexity
- Month-long release cycles
- Hidden operational overhead
- Cumulative maintenance debt

The worst consequence is not even cost, it's **loss of control**.

Once you depend on a dozen subsystems, your architecture no longer belongs to you.

It belongs to your integrations.

AI HAS ACCELERATED THE CRISIS

The rise of AI and vector-based workloads has transformed data into a much more multi-model asset:

- structured objects
- documents
- embeddings
- metadata
- files
- binary assets
- relationships
- logs
- audit trails
- semantic neighborhoods

To support this, teams bolt on a vector database, then a graph engine, then an embedding pipeline, then new forms of indexing.

But with each addition, the architecture fractures further.

AI demands unified context.

But our systems are becoming more disjointed than ever.

This tension is now so sharp that many teams feel paralyzed: they cannot evolve their product without expanding an already unstable stack.

The complexity crisis isn't an abstraction.

It is now a daily operational reality.

THE PRIVACY AND COMPLIANCE WALL

While AI pulls architectures apart, another force pushes in the opposite direction: **the global privacy and compliance wave.**

- GDPR
- CCPA
- NIS2
- DGA
- HIPAA
- SOC2
- PCI
- ISO 27001
- CSF 2.0

Regulatory frameworks expect:

- full provenance
- tamper-proof audit trails
- field-level access controls
- encryption everywhere
- tenant isolation
- explainable data flows
- selective deletion
- deterministic compliance boundaries

But the moment a system spans 10+ interconnected components, these requirements become nearly impossible to satisfy reliably.

Security models drift. Audit guarantees vary. Tenancy boundaries weaken. Validation lives in multiple places. Deletion becomes probabilistic.

Most architectures survive only through brute force: patches, glue code, scripts, ad hoc processes, and hope.

THE UNSPOKEN TRUTH

The industry has been quietly adapting around a deep structural flaw:

> **The data architecture we inherited was never designed for the systems we are building today.**

Databases were built for structured tables.

Document stores were built for semi-structured JSON blobs.

Search engines were built for text.

Vector stores were built for embeddings.

Graph databases were built for traversals.

Object stores were built for large files.

Each was built with a different model of:

- concurrency
- indexing
- security
- authorization
- schema
- identity
- provenance
- tenancy
- transactions

And we are left trying to make them behave like one coherent system.

It doesn't work.

It's not scalable.

And it will only get worse.

A NEW PARADIGM IS NEEDED

What the industry lacks is not:

- a better database,
- or a faster vector engine,
- or a more expressive graph model,
- or a more scalable document store.

What we lack is a **unified architecture** that brings these capabilities together under a **single schema, single execution model, single security model, single provenance trail, and single mental model**.

This is the fundamental insight that leads to the Hyperstore concept:

Databases are too small.

Data platforms are too fragmented.

What we need is something new.

Something built for:

- AI-native systems
- multi-model data
- privacy-by-design
- deterministic generation
- multi-tenant architectures
- secure collaboration
- continuous compliance
- zero-trust environments

Something that replaces the stack tax with a coherent architecture.

Something that eliminates the division between object, vector, graph, and file storage.

Something that finally unifies the backend.

This is the birth of the **Hyperstore**.

And it is the problem *Vektagraf* was built to solve.

In the next chapter, we examine why this crisis cannot be resolved by incremental improvements to existing databases, and why the limitations are structural rather than operational.

CHAPTER 2
WHY TRADITIONAL DATABASES CAN'T EVOLVE FAST ENOUGH

The modern backend stack is a museum of data technologies, each representing an era when its particular capability was the missing piece. Relational databases solved structured storage. Document databases solved flexible schemas. Graph databases solved relationships. Search engines solved indexing. Vector databases emerged to support AI. And file/object stores handled the rest.

Each system emerged to address a legitimate need. None of them were designed to work together. All of them are now expected to.

The result is a paradox of progress: the more problems we solve, the more fragmented the landscape becomes. And as the data ecosystem grows, the gap widens between what modern applications require and what traditional databases were ever meant to do.

This is not a failure of engineering, it is a mismatch of eras.

THE RELATIONAL PLATEAU

Relational databases have endured for decades because their core abstraction, tables with strongly typed schema, provides predictability, consistency, and a simple conceptual model. But this same rigidity becomes a trap in modern workloads.

Relational systems struggle when:

- the shape of data evolves frequently,
- hierarchical or nested objects become the norm,
- versioning is required,
- multilingual content expands alongside structured fields,
- unstructured documents appear next to structured records,
- relationships form complex networks,
- embeddings need to be co-located with metadata,
- full-text and vector search must be used simultaneously.

Engineers have spent years trying to extend relational databases to fit these needs, but these extensions often violate the foundational assumptions that make relational systems reliable in the first place.

The relational model is not broken.

It is simply complete, finished; unable to stretch far enough to occupy the new terrain.

DOCUMENT STORES: FLEXIBILITY THAT CUTS BOTH WAYS

Document databases emerged as a reaction to the rigidity of SQL schemas. They promised flexibility, developer friendliness, and rapid iteration. And they delivered, initially.

But in exchange for schema freedom, organizations absorbed new forms of chaos:

- silent schema drift,
- inconsistent field types,
- validation only at the edge (if at all),
- difficulty enforcing invariants,
- challenges with multi-document transactions,
- deeply inconsistent documents across tenants, versions, and services.

Over time, companies reintroduced schemas, validation, and type systems on top of their document stores, reinventing much of what relational databases had already mastered.

The flexibility that once felt liberating now frequently becomes a liability.

Document stores were never meant to unify multiple data modalities, enforce principled constraints, or participate meaningfully in privacy-by-design architectures. Their freedom is structural, not intentional.

SEARCH ENGINES: A SEPARATE WORLD WITH SEPARATE TRUTHS

Search engines are powerful, but they operate on an entirely different worldview:

- data is copied,
- transformed,
- inverted,
- sharded,
- and indexed using text-centric algorithms.

They maintain a **second truth**, parallel to the database.

Synchronization is fragile and frequently stale.

Security and tenancy models rarely align with the primary system.

The gap between "stored data" and "searchable data" expands with each new feature:

- fuzzy search,
- scoring models,
- semantic indexing,
- embeddings,
- hybrid search.

Every new capability deepens the rift.

Search engines were never intended to be authoritative data stores, and so the more we rely on them, the more architectural incoherence we inherit.

VECTOR DATABASES: THE NEWCOMER WITH A NARROW WORLDVIEW

The rapid emergence of vector databases reflects a profound trend: AI workloads demand storage strategies that legacy systems simply can't support.

But today's vector databases are highly specialized:

- They optimize for similarity search, not broad data modeling.
- They treat embeddings as stand-alone entities.
- They rarely integrate strongly typed fields, validation, or constraints.
- They lack fine-grained tenancy, provenance, or compliance guarantees.

- They store relationships only implicitly as cosine distances, not explicit graph semantics.
- They cannot serve as authoritative systems of record.

As a result, every AI team ends up maintaining:

- a primary database,
- a vector store,
- a search engine,
- and some glue logic for synchronizing them.

Vector databases solved an urgent problem, but they also multiplied infrastructure complexity.

They were never meant to be general-purpose data systems.

Nor should they be.

GRAPH DATABASES: GREAT FOR RELATIONSHIPS, BAD FOR EVERYTHING ELSE

Graph databases excel at modeling connections. But their strengths are hyper-specific, and attempts to broaden their scope have not meaningfully changed their fundamental limitations:

- They struggle with large binary assets,
- They cannot efficiently handle document-like flexibility,
- They cannot represent dense embeddings as first-class citizens,
- They provide limited multi-tenancy support,
- They rely on specialized query languages incompatible with the rest of the stack.

Organizations that adopt a graph database rarely replace their SQL or document system, they **add** a graph database.

Another component.

Another mental model.

Another operational dependency.

Graph databases were optimized for relationships, not unification.

OBJECT STORES: THE LAST RESORT STORAGE LAYER

Object stores, S3 and its many derivatives, are the duct tape of modern architectures. Anything the database can't handle ends up here:

- images
- video
- blobs
- archives
- binary payloads
- embeddings stored as files
- serialized graphs
- exported state
- machine learning artifacts

Object stores are not databases.

They do not validate.

They do not guarantee consistency.

They do not index.

They do not enforce constraints.

They do not understand structure.

There is no schema.

There is no provenance.

There is no tenant isolation.

There is no audit trail.

This is where the "unknown parts of the system" accumulate.

Every object store becomes a graveyard of context.

POLYGLOT PERSISTENCE: A BEAUTIFUL IDEA THAT DOESN'T WORK IN PRACTICE

The industry once believed that multiple specialized databases could be orchestrated into a harmonious architecture. This idea, **polyglot persistence**, was academically appealing but operationally disastrous.

The promise:

> Use the best database for each workload.

The reality:

> Now you have six databases, none of which agree with each other.

Polyglot systems fail because:

- Their schemas diverge over time
- Their security models contradict each other
- Their access controls are incompatible
- Their indexing strategies differ
- Their transaction models cannot be unified
- Their audit trails do not correlate
- Their backup strategies don't align

- Their privacy guarantees are inconsistent
- Their performance characteristics clash
- Their APIs are incompatible
- Their provenance disappears across boundaries

When an architecture is composed of fundamentally different systems, **consistency becomes optional and correctness becomes accidental**.

THE HARD LIMIT: DATABASES CANNOT EVOLVE FAST ENOUGH

Each generation of data system was optimized for a specific era:

- SQL for structured business data
- Document stores for web-scale JSON
- Search engines for text indexing
- Vector databases for embeddings and AI
- Graph databases for complex relationship traversal
- Object stores for unstructured payloads

None of these systems were designed to:

- unify modalities,
- enforce consistent privacy,
- validate all data at ingest,
- express provenance natively,
- support multi-tenancy as a core concept,
- provide AI-native query semantics,
- govern schema evolution across an entire backend,
- generate entire SDKs and APIs,
- combine vectors + objects + graphs in a single query.

Databases are evolving, but the world is evolving faster.

They are trying to patch over architectural assumptions that are now decades old.

> **You cannot retrofit unification onto systems built for fragmentation.**

THE ARCHITECTURAL MISMATCH HAS BECOME TERMINAL

What we see today is not a failure of databases.

It is a **failure of scope**.

Traditional data systems were meant to be pieces of larger architectures. But modern systems demand unification at a level that those tools were never built to support. The gap between requirements and capability is widening, not shrinking.

And every attempt to bridge that gap, plugins, adapters, wrappers, ETL pipelines; further exposes the underlying mismatch.

This is why the future cannot be "a better database."

It must be a different category entirely.

The world needs an architecture that:

- natively unifies objects, vectors, graphs, and files
- enforces privacy as a core constraint
- guarantees provenance and auditability
- supports multi-tenant isolation at every layer
- integrates validation and automation
- maintains a single schema as a single source of truth
- generates its own APIs and SDKs
- provides one security model
- provides one mental model

- and eliminates the stack tax permanently

The industry has spent years optimizing pieces.

It is time to optimize the whole.

This is the inflection point that leads directly to the **Hyperstore**.

In the next chapter, we define the concrete requirements modern systems now impose on data architectures, requirements that existing categories were never designed to meet.

CHAPTER 3
THE NEW REQUIREMENTS OF MODERN SYSTEMS

The demands placed on modern software systems have evolved faster than the architectures supporting them. What once passed for a "scalable backend" is now inadequate for applications that must operate across multiple data modalities, enforce strict privacy guarantees, support AI workflows, and serve global multi-tenant environments.

We are no longer building simple CRUD applications.

We are building **knowledge systems**, **AI-driven products**, **compliance-dependent platforms**, and **zero-trust architectures**.

The old stack cannot keep up.

This chapter sets the stage for why a new category — **the Hyperstore** — must exist.

MULTI-MODEL DATA AS THE NEW DEFAULT

Historically, systems revolved around structured records. But modern applications must integrate many forms of data simultaneously:

- structured objects
- documents and nested structures
- graphs and relationships
- vector embeddings
- raw text
- binary files and media
- logs, traces, and event streams
- internationalized content
- cryptographic signatures
- provenance metadata

These are not separate concerns.

They are *the same entity* expressed through different modalities.

Example: A user profile may now include:

- personal info (structured)
- permissions (graph)
- preferences (embedding)
- activity logs (event stream)
- uploaded content (files)
- translations (i18n content)
- cryptographic proofs (binary)

Traditional architectures split this across multiple systems.

But the application needs *one* unified representation.

Modern systems must treat multi-model data as atomic and inseparable.

AI-NATIVE WORKFLOWS REQUIRE DEEP DATA UNIFICATION

AI does not respect the boundaries our systems impose.

A model embedding a document needs:

- the document text,
- metadata,
- relationships,
- previous versions,
- permissions,
- user context,
- tenant context,
- audit logs,
- sometimes binary attachments.

In legacy stacks, these are stored across:

- database tables
- document stores
- search clusters
- vector databases
- object storage
- log pipelines
- policy engines
- user directories

This fragmentation fundamentally limits what AI can do.

To make AI an operational capability, not just an add-on, systems must:

- co-locate embeddings with structured fields
- relate embeddings to graph edges

- search across text, vectors, and metadata
- enforce security and tenant boundaries in inference
- track provenance of model inputs
- validate that embeddings reflect current data
- support vector updates tied to domain events

In short:

AI is only as powerful as the architecture that feeds it.

Legacy architectures cannot provide the necessary cohesion.

PRIVACY-BY-DESIGN IS NO LONGER OPTIONAL

Global regulatory frameworks have shifted the burden of trust directly onto system design. It is no longer acceptable to bolt security on after development.

The law expects:

- complete provenance
- explicit lineage tracking
- auditability of every change
- multi-tenant guarantees
- encryption everywhere
- field-level access boundaries
- the ability to delete *precisely and permanently*
- minimization of stored data
- traceable data flows
- isolation of cryptographic identity
- deterministic behavior under compliance review

This is not achievable in fragmented stacks.

In practice, most companies cannot answer:

- "Where did this value come from?"
- "Who had access to this field at what time?"
- "Was this metadata generated before or after permission changed?"
- "Can we prove this vector embedding corresponds to this document version?"
- "If a user requests erasure, can we guarantee *all* derived data is removed?"

A system of loosely connected databases cannot provide deterministic answers.

A Hyperstore must.

POST-QUANTUM AND ADVANCED CRYPTOGRAPHY REQUIREMENTS

The next decade introduces cryptographic requirements unprecedented in mainstream systems:

- **Post-quantum key exchange** (e.g., Kyber)
- **Post-quantum digital signatures** (e.g., Dilithium)
- **Homomorphic encryption** for private computation (e.g. Pallier)
- **Differential privacy** for aggregate insights
- **Selective disclosure** of fields
- **End-to-end cryptographic integrity**
- **Encrypted provenance**
- **Multi-party computation primitives**

Traditional databases were never designed to:

- store encrypted-by-design fields,
- compute with encrypted numbers,
- tie identity to cryptographic proofs,

- maintain post-quantum signature chains,
- ensure consistency between encrypted and unencrypted representations.

These capabilities are now critical for regulated industries, data-sensitive products, and long-lived systems.

Architectures must treat cryptography as a first-class concern, not an extension.

MULTI-TENANT SAAS HAS BECOME THE BASELINE

Multi-tenancy is no longer a special case, it's the foundation of modern Software as a Service (SaaS).

Tenancy must be:

- enforced in the data model
- guaranteed by the storage engine
- reflected in the schema
- integrated with identity
- validated across queries
- enforced in vectors, objects, files, and relationships
- auditable
- cryptographically provable

Traditional systems model tenancy at the application layer.

This is fragile, error-prone, and nearly impossible to audit.

Misconfiguration becomes breach.

A single incorrect query becomes a compliance incident.

Modern systems require:

- tenant-aware storage

- tenant-separated indexes
- tenant-scoped embeddings
- tenant-bound provenance
- tenant-specific automations
- tenant-limited workflows

The data layer, not the application, must enforce boundaries.

PROVENANCE AND AUDITABILITY ARE NOW FUNCTIONAL REQUIREMENTS

The rise of AI, compliance, and distributed collaboration makes provenance essential.

Organizations must answer:

- What changed?
- When?
- Who caused it?
- What was the previous value?
- Was the value encrypted?
- Was a workflow triggered?
- Was the tenant boundary respected?
- Is the data still consistent with its embedding?
- Can we reproduce a historical query result?

Traditional databases can log changes, but they cannot:

- capture semantic intent,
- store cryptographic signatures,
- relate provenance to objects, vectors, and files,
- audit field-level access,
- enforce version-level consistency,
- rebuild historical states safely,
- validate lineage end-to-end.

A modern platform must treat provenance as a *governance substrate,* not a log.

DETERMINISTIC BACKEND GENERATION IS BECOMING NECESSARY

Building backends manually is becoming untenable:

- Data models evolve continuously
- APIs must remain synchronized
- Validation must be consistent across layers
- Clients require typed, predictable contracts
- Migrations must be safe and deterministic
- Access control must be uniform
- Tenancy and policy boundaries must be enforced consistently

When systems are hand-built, model drift is inevitable.

Developers maintain:

- models in the app
- models in the schema
- models in the API
- models in the client
- models in the database
- migrations in separate code
- access rules in multiple layers

Any divergence creates silent risk.

Backends must become **generated, not written**.

This requires a unified schema model, something traditional databases were never designed to provide.

ZERO-TRUST ARCHITECTURE REQUIRES DATA-LEVEL AUTHORITY

Zero-trust principles demand:

- no implicit trust
- no perimeter reliance
- verification at every step
- identity-rooted access
- data-level enforcement of secrets
- auditability and revocation
- cryptographic guarantees

Legacy architectures enforce these constraints:

- in code,
- in middleware,
- in API gateways,
- in IAM integrations,
- in front-end logic,
- across multiple databases.

This creates fragility.

The security model must live **inside the data architecture**, not outside it.

UNIFIED CONTEXT IS THE NEW COMPETITIVE ADVANTAGE

Applications that understand their data deeply — *semantically, structurally, historically* — win.

Systems that scatter context across multiple databases cannot support:

- AI agents
- real-time recommendations
- cross-model queries
- end-to-end personalization
- advanced search
- contextual automation
- context-enriched workflows

A system that unifies all modalities can.

THE REQUIREMENTS ARE CLEAR

The modern system must support:

- **Multi-model data as one cohesive model**
- **AI-native semantics**
- **Privacy-by-design and post-quantum security**
- **Multi-tenancy built into the storage layer**
- **First-class provenance and auditability**
- **Deterministic backend generation**
- **Unified context for search, analytics, and intelligence**
- **Validation, automations, and governance as built-in capabilities**
- **Zero-trust guarantees**
- **Elastic performance and storage scalability**

None of these requirements can be met by stitching together existing tools.

They demand an architecture built for unification.

They demand a ***Hyperstore***.

In the next chapter, we return to the developer, examining how these unmet requirements manifest as cognitive overload, architectural fragility, and systemic risk.

CHAPTER 4
THE DEVELOPER'S DILEMMA, REVISITED

Developers are the first to feel architectural pain, often long before an organization realizes the depth of the problem. They are the ones writing glue code between services, debugging mismatched schemas, reconciling divergent data representations, stitching search indexes to primary stores, and maintaining complex pipelines just to keep the system coherent.

The "Developer's Dilemma" is no longer just about choosing the right tool for the job. It is about surviving a landscape where every choice creates more complexity downstream. This dilemma is now structural, systemic, and increasingly unsustainable.

Modern backend engineering has turned into a negotiation between competing constraints rather than a creative act of building.

THE COGNITIVE OVERLOAD OF FRAGMENTED ARCHITECTURES

Every system added to the stack becomes another mental model a developer must carry:

- a different query language
- a different type system
- a different security model
- a different API surface
- a different index strategy
- a different concurrency behavior
- a different failure mode

Developers are expected to weld these models into something unified for the application. But that unification exists only in their heads and in ad hoc code, they are compensating for architectural incoherence that the underlying systems do not provide.

The more the stack grows, the more developers become interpreters of conflicting truths.

This cognitive burden has real consequences:

- slower development
- unpredictable bugs
- fragile integrations
- inconsistent behavior across services
- difficulty onboarding new engineers
- a constant feeling of "swimming upstream"

When the system is fragmented, the developer becomes the glue.

And humans are the *least reliable* part of the stack.

THE FALSE CHOICE BETWEEN FLEXIBILITY AND SAFETY

Developers face a perennial dilemma:

> Choose flexible tools (document databases, object stores)...

> Risk schema drift, inconsistent data, weak validation.

> Choose rigid tools (relational databases)...

> Low iteration, migration headaches, reduced agility.

Modern products require both flexibility and safety.

Traditional databases force a trade-off.

Developers must choose:

- Iterate quickly and risk chaos
- Or enforce structure and slow delivery

This tension leaves teams constantly oscillating between agility and control, often landing in a middle ground where neither is fully achieved.

A Hyperstore must unify these two extremes:

- flexible modeling for iteration
- strongly enforced constraints for correctness

Without forcing developers to choose one or the other.

SCHEMA DRIFT: THE INVISIBLE ENEMY

In fragmented architectures, schemas do not drift, they **diverge**.

Developers maintain:

- schema in the application
- schema in the database
- validation in the API
- types in the client
- indexing rules in the search engine
- embeddings in a vector store
- object metadata in an object store
- permissions elsewhere
- provenance outside the database
- analytics schema in a warehouse

No two remain perfectly aligned.

This divergence creates consequences developers deal with daily:

- fields that exist only "somewhere"
- inconsistent data formats
- code paths that assume different shapes
- vector embeddings that don't match updated documents
- search indexes missing fields
- analytic pipelines failing silently
- invalid data silently entering the system

Developers become custodians of schema correctness across dozens of systems.

This is not their job.

It's a sign of architectural decay.

A Hyperstore must eliminate schema drift entirely, by making the schema the authoritative, singular source of all logic.

THE FRAGILITY OF DISTRIBUTED GOVERNANCE

In a multi-system architecture:

- validation lives in some services
- access control lives in other services
- tenancy restrictions live in the application layer
- security rules are scattered
- audit logs exist outside the datastore
- provenance is applied inconsistently
- workflows run in external orchestrators

Developers must manually reconcile these governance paths.

A missing check, forgotten rule, or misapplied filter becomes:

- a privacy breach
- a cross-tenant data leak
- a silent validation failure
- corrupted embeddings
- incorrect graph relationships
- incomplete audit trails
- a compliance incident waiting to happen

Governance that is **distributed** is governance that is **fragile**.

Developers shouldn't carry this burden.

The architecture should.

THE GLUE CODE TRAP

When the architecture does not unify modalities, developers write the unifying logic themselves.

Glue code is everywhere:

- listeners syncing data into search indexes
- ETL jobs updating vector embeddings
- scripts rehydrating objects from object stores
- ad hoc validation utilities
- custom migration scripts
- wrappers around access control
- custom provenance logic

Glue code:

- is invisible in diagrams
- is brittle under change
- is rarely tested thoroughly
- becomes tribal knowledge
- tends to break at inconvenient times
- hides systemic complexity behind developer labor

The system does not survive because it is well-designed, it survives because developers keep it alive.

This is a form of **structural technical debt**.

And it grows every day.

THE BUILD-VS-SCALE PARADOX

Developers frequently face a painful paradox:

Build fast using flexible tools → rearchitect later.

Build with scalable architecture → slow initial development.

This dilemma is not caused by developer choices.

It is caused by architectural constraints.

A system that is easy to build is often hard to scale.

A system that is hard to build is often easy to scale.

This reinforces a myth:

"You must choose between building fast and building right."

A Hyperstore aims to abolish this trade-off entirely, providing an architecture that is both fast to build on and inherently scalable.

AI WORKFLOWS CREATE UNMANAGEABLE DEVELOPER BURDEN

Developers are now expected to integrate:

- model inference
- embeddings
- semantic search
- vector updates
- model-driven automations
- scoring functions
- hybrid search between structured and vector queries
- embeddings that must be tied to versioned data
- cross-model joins

Each part lives in a different system.

Developers must manually orchestrate it.

This makes AI integration:

- brittle
- expensive
- slow to evolve
- prone to inconsistencies
- difficult to govern
- impossible to audit end-to-end

A Hyperstore treats vectors, objects, and graphs the same way,

eliminating entire classes of glue code and validation logic.

DEVELOPERS ARE CARRYING THE ARCHITECTURE ON THEIR SHOULDERS

When the architecture is fragmented, complexity flows downward to the developer.

They become responsible for:

- keeping systems in sync
- translating between modalities
- enforcing constraints across boundaries
- maintaining scattered security rules
- writing documentation that explains the system's inconsistencies
- building migrations for every change
- debugging failures that originate in another system
- patching misalignments in data semantics
- understanding multiple query languages
- maintaining mental models no single tool provides

Developers have effectively become the integrators-of-last-resort.

This is not sustainable.

It is not efficient.

And it is not the right use of human talent.

An architecture should support developers, not rely on them to compensate for its weaknesses.

THE NEED FOR A UNIFYING ARCHITECTURE

Developers are signaling, loudly and continuously, what the industry refuses to hear —

We need an architecture that:

- reduces mental load
- unifies modalities
- enforces schema consistency
- integrates privacy and governance
- collapses redundant infrastructure
- generates backend logic deterministically
- provides a single place for rules, automations, and constraints
- makes multi-tenancy safe by construction
- enables AI without stitching multiple systems together

We need an architecture that isn't assembled, but **designed**.

This is the developer's dilemma:

They are forced to build coherent systems out of incoherent parts.

The solution is not better documentation, better glue code, or better orchestration.

The solution is **a unified data architecture** that eliminates fragmentation.

That architecture is the *Hyperstore.*

In the next part, we introduce the Hyperstore as a new architectural category designed to resolve these tensions at their source.

PART II

INTRODUCING THE NEW CATEGORY

For decades, the data landscape has expanded horizontally: new databases, new indexers, new formats, new stores, each addressing a specific need, none addressing the whole. What modern systems require is *vertical integration* at the data layer: a unified architecture that treats every modality, every rule, every workflow, and every tenant boundary as facets of the same underlying truth.

This is the Hyperstore.

A Hyperstore is not a database with extra features, nor is it a fusion of multiple existing systems. It is a fundamentally new category of data architecture designed to unify everything a modern application needs into a single, coherent platform.

If the database was the foundation of the last generation, **the Hyperstore is the foundation of the next.**

A system cannot guarantee what it does not control.

CHAPTER 5
THE HYPERSTORE INVENTION

A Hyperstore is a unified, multi-model, self-governing data platform that integrates storage, indexing, validation, automation, security, provenance, and generated APIs into one cohesive architecture.

It is not "one database to rule them all."

It is **one architecture to eliminate the need for many.**

A Hyperstore provides:

- **A single schema** describing *all* data, rules, workflows, access models, and types
- **A unified storage engine** for objects, graphs, vectors, and files
- **Integrated validation** at the data boundary
- **Embedded governance** for privacy, multi-tenancy, rules, provenance
- **Built-in automation** triggered by data events
- **Deterministic code generation** for clients and servers
- **Unified indexing** for structured, text, and semantic search

- **Cryptographic foundations**, including post-quantum support
- **Zero-trust data semantics** by design
- **AI-native representations** (vectors, embeddings, metadata alignment)

No separate search index.

No separate vector store.

No separate object store.

No separate workflow engine.

No separate schema registry.

No separate governance layer.

No separate code generators.

Everything derives from one source of truth.

THE CORE PREMISE: UNIFICATION, NOT AGGREGATION

A Hyperstore is not built by stitching systems together, it is built by designing each part of the architecture to coexist natively with every other part.

This means:

- **Vectors** understand and reference **objects**, **files**, and **graph relationships**
- **Objects** can embed structured fields, text fields, binary content, relationships, and vector embeddings
- **Graph edges** link across all modalities
- **Workflows** introspect provenance and policy
- **Validation** is applied uniformly everywhere

- **Search** spans structured, unstructured, and semantic fields in one operation
- **Tenancy** is enforced at every level of the architecture
- **Encryption** applies to both data and metadata structures
- **Provenance** is intrinsic to all updates

Hyperstores are **vertically integrated** systems,

where the layers are not separate technologies but expressions of the same data model.

THE HYPERSTORE STACK MODEL

A Hyperstore contains several integrated layers that together form a vertically unified data platform:

Data Layer

- Structured objects
- Documents and nested fields
- Graph relationships
- Vector embeddings
- File and binary storage

Indexing Layer

- Structured indexes
- Full-text indexes
- Vector and semantic indexes
- Combined hybrid search operations

Governance Layer

- Validation and invariants
- Internationalization rules

- Access control and tenancy
- Provenance and auditability
- Policy enforcement
- Field-level privacy semantics

Execution Layer

- Automations
- Triggers
- Background jobs
- Derived fields
- Event-driven workflows

Cryptography Layer

- Encryption at rest and in-flight
- Per-field encryption
- Differential privacy
- Post-quantum signature chains
- Homomorphic computation
- Integrity verification

Generation Layer

- Server-side code generation
- Client SDK generation
- Type-safe models
- Query utilities
- Migration logic
- Endpoint definitions

Unlike traditional systems, these layers do not bolt onto each other, they reflect one schema and operate as one system.

THE HYPERSTORE SCHEMA

In a Hyperstore, the schema is the single source of truth, defining:

- models
- fields
- types
- relations
- collections
- indexes
- validations
- privacy rules
- tenant boundaries
- internationalization
- workflow triggers
- automations
- provenance policies
- cryptographic requirements

The schema does not contain executable code.

It acts as a *formal contract* that the Hyperstore interprets and generates code from.

This means:

- backend APIs are generated deterministically
- client SDKs always match the backend
- migrations are safety-checked
- validation is guaranteed to be consistent
- policies cannot drift
- model definitions cannot diverge
- documentation is always in sync

The schema is not an artifact.

It is the architecture.

THE HYPERSTORE QUERY MODEL

Hyperstores unify:

- structured queries
- graph traversal
- text search
- vector similarity
- hybrid search
- file metadata queries

This produces a radical shift:

Without switching tools, all modalities participate in one query ecosystem.

This unlocks capabilities like:

> Find all documents related to this concept, written in French, attached to users with a similar activity pattern, uploaded in the last 30 days.

> Retrieve customer records whose embeddings suggest churn risk, then traverse their relationship graph to identify upstream dependencies.

> Search across text, metadata, and vector similarity in a single operation.

Hyperstore queries are multi-model, multi-index, and contextually aware.

SELF-GOVERNANCE: THE HALLMARK OF A HYPERSTORE

A defining trait of the Hyperstore is **self-governance**.

Instead of pushing validation, permissions, rules, and workflows to application code, the Hyperstore enforces them at the data layer:

- Validation that cannot be bypassed
- Access control that cannot be circumvented
- Automations that fire reliably
- Tenancy boundaries that cannot be misconfigured
- Provenance that cannot be tampered with
- Encryption that cannot be forgotten
- Policies that cannot drift

Your application cannot "forget" to enforce the rules.

The Hyperstore *is* the enforcement.

This drastically reduces systemic risk and developer burden.

THE HYPERSTORE PERFORMANCE MODEL

Traditional architectures scale by distributing separate systems.

Hyperstores scale by distributing **one unified system**.

This enables:

- shared caching
- unified memory models
- consistent concurrency semantics
- predictable transactional behavior
- efficient index alignment
- tightly coupled vector + structured indexing

- optimized access paths that cross modalities

Hyperstores enhance performance not by selecting specialized engines, but by eliminating the overhead caused by stitching specialized engines together.

The result is a simpler, faster, more consistent architecture.

THE HYPERSTORE: A NEW TRUST FOUNDATION

In AI-driven applications, global privacy compliance, zero-trust security expectations, and multi-tenant SaaS, trust can no longer be an external system. It must be an intrinsic property of the data architecture.

The Hyperstore becomes a trust substrate:

- semantic correctness
- consistent rules
- guaranteed provenance
- auditable lineage
- verifiable identities
- cryptographic integrity
- tenant isolation
- deterministic behavior

Trust is no longer a configuration.

Trust is the architecture.

WHY A HYPERSTORE IS NOT "JUST A BETTER DATABASE"

A Hyperstore is not a database with extra capabilities.

It is a deeply reimagined architecture that:

- eliminates the need for multiple databases
- redefines how backends are built
- makes governance a built-in concept
- unifies modalities without glue code
- collapses entire infrastructure layers
- produces deterministic application backends
- introduces a model where data, rules, and workflows coexist seamlessly

It is not an incremental evolution.

It is a categorical shift.

THE HYPERSTORE IN CONTEXT

If databases defined the last 40 years, Hyperstores will define the next 40.

They answer the demands of modern systems:

- AI-native
- multi-model
- privacy-centric
- governance-enforced
- multi-tenant
- developer-friendly
- zero-trust
- unified
- generative
- cryptographically secure

And they collapse the fragmentation that the industry has normalized.

CONCLUSION

A Hyperstore is the architecture the modern era demands, the only architecture capable of unifying data, logic, governance, security, and application generation into a single coherent whole.

Now that we understand what a Hyperstore is, the next chapter explains how Hyperstores reimagine the architecture of modern systems, and why this approach wasn't feasible until now.

CHAPTER 6
WHY HYPERSTORES ARE ARCHITECTURALLY POSSIBLE NOW

(AND NOT EARLIER)

If Hyperstores feel inevitable today, it is because the industry has finally converged on the technological, conceptual, and infrastructural foundations required to make them real. For decades, the idea of an "everything store for data", one engine capable of unifying objects, graphs, vectors, files, workflow logic, privacy guarantees, provenance, cryptography, and multi-tenancy; was unworkably ambitious.

Hyperstores were impossible not because the industry lacked imagination, but because **the prerequisites did not yet exist**.

Only in the mid-2020s did enough trends converge to make the architecture feasible. This chapter examines the underlying forces that opened the door, and why this moment is uniquely suited for the emergence of the Hyperstore.

THE RISE OF JSON AND UNIVERSAL STRUCTURED DATA

A Hyperstore relies on a universal, expressive, typed object model.

This only became possible when **JSON became the lingua franca of data interchange**.

Between 2010 and 2020:

- Nearly all programming languages adopted rich JSON support
- APIs standardized around structured JSON payloads
- Schemas evolved (OpenAPI, JSON Schema, GraphQL type systems)
- Cloud services natively supported JSON-oriented metadata
- Databases began to accept JSON as a first-class field type

This created an unprecedented alignment across languages, services, and systems. For the first time in history, nearly every tool and platform could represent the same object model.

This alignment provided the base material for a unified storage architecture. Without this, a Hyperstore could not have exposed a single shared schema model across all systems.

THE MATURATION OF STRONGLY TYPED SCHEMA LANGUAGES

While JSON provided the ecosystem-wide substrate, the last decade also saw the widespread adoption of higher-level type systems:

- TypeScript
- GraphQL SDL
- JSON Schema
- Protobufs
- domain-specific schema engines

The industry shifted from runtime validation to **compile-time, schema-driven correctness**. By the early 2020s, developers were accustomed to the idea that:

- backend and frontend types should match,
- APIs should be generated from schema,
- consistent types reduce systemic risk,
- schema-first development improves quality.

Hyperstores rely on a single deterministic schema to generate all layers of the stack.

This mindset simply didn't exist 10–15 years ago.

CLOUD-NATIVE INFRASTRUCTURE MADE VERTICAL INTEGRATION FEASIBLE

Hyperstores achieve unification by integrating multiple subsystems, storage, indexing, validation, workflow execution, search, and cryptography; into one architecture.

This requires:

- elastic storage
- distributed concurrency
- shared caching layers
- global replication
- containerized environments
- standardized runtime expectations
- predictable latency characteristics

Only after cloud computing became dominant did it become possible to run a vertically integrated data architecture that could:

- scale horizontally on commodity infrastructure
- coordinate shared indexes
- isolate thousands of tenants
- replicate provenance logs globally
- unify memory strategies across modalities

Hyperstores fundamentally depend on cloud-era assumptions. They are cloud-native architectures that could not have existed in a world of monolithic, single-node databases.

THE EXPLOSION OF AI AND VECTOR EMBEDDINGS

Before the AI revolution, embeddings were an academic curiosity, not a fundamental design requirement.

Modern applications now require:

- vector similarity search
- hybrid search (metadata + vector + text)
- embedded semantic metadata
- contextual models
- entity-linked embeddings tied to object graphs
- consistent updates across modalities

Most traditional databases cannot integrate vectors natively.

Vector databases emerged, but brought fragmentation.

By 2023–2025, it became clear that:

- AI requires unified data.
- Fragmentation breaks context.

- Glue code breaks consistency.

The AI boom forced the industry to confront the limits of its architectures.

Hyperstores are a response to this pressure: **a system designed for AI-native, multi-model, unified representations.**

This requirement simply didn't exist before.

PRIVACY, GOVERNANCE & COMPLIANCE BECAME CORE SYSTEM REQUIREMENTS

GDPR (2018) started the wave.

CCPA, CPRA, HIPAA modernization, PSD2, SOC expansions, and AI governance frameworks followed.

By the late 2020s, companies were expected to:

- track full data lineage
- enforce tenant isolation
- prove access decisions
- demonstrate schema-level intent
- delete derived/embedded data correctly
- provide audit trails for AI inputs
- apply privacy rules consistently
- secure data using modern cryptography

Traditional databases cannot natively support:

- field-level provenance
- differential privacy
- post-quantum signatures
- tenant-bound indexing
- complete lineage reconstruction

Before this regulatory shift, Hyperstores weren't economically justified.

Compliance pressure made them necessary.

THE ZERO-TRUST SECURITY MOVEMENT

The shift from network trust to identity-verified, cryptographic trust changed expectations permanently.

Zero-trust requires:

- every access verified
- every data boundary explicit
- every tenant isolated
- every update auditable
- every rule enforced at the lowest possible layer

This drove the industry toward architectures where **security lives inside the data layer**, not in the application wrapper.

Hyperstores embrace this shift: the data model itself encodes security.

This was not technically or culturally feasible in earlier eras.

ADVANCES IN APPLIED CRYPTOGRAPHY

The idea of integrating post-quantum and advanced cryptography directly into a datastore would have been absurd in the 2000s and impractical in the early 2010s.

By the mid-2020s:

- Kyber and Dilithium were standardized
- differential privacy libraries matured
- homomorphic encryption became performant enough for limited domains

- privacy budgets were better understood and measurable
- trusted execution environments proliferated
- key management as a service became ubiquitous

This new cryptographic landscape made it possible to embed:

- encrypted computation
- encrypted provenance
- post-quantum key exchange
- cryptographically bound tenancy
- signed object lifecycles

Hyperstores depend on these new primitives.

Without them, the trust substrate would be impossible.

DEVELOPER EXPECTATIONS HAVE EVOLVED BEYOND DATABASES

Developers now assume:

- client SDKs should be generated
- backend APIs should be generated
- schema should define the world
- migrations should be deterministic
- data modeling should be declarative
- governance should not rely on application code
- privacy should be enforced at the data layer
- multi-modality should be native
- AI should be integrated, not bolted on

This cultural shift created demand for a new class of system.

Earlier generations of developers were database-centric.

Today's developers are **schema-centric**.

Hyperstores are a response to this shift in expectations.

DISTRIBUTED SYSTEMS THEORY FINALLY STABILIZED

Between 2010 and 2020, the industry learned (often painfully) how to operate:

- eventual vs strong consistency
- multi-version concurrency control
- distributed logs and WAL replication
- conflict-free replicated data types (CRDTs)
- sharding strategies
- garbage collection in distributed stores
- low-latency consensus algorithms
- vector clocks and lineage models

This body of knowledge made it possible to design systems that are:

- multi-tenant
- distributed
- consistent
- governable
- performant
- auditable

Earlier eras lacked the theoretical maturity to support architecture as ambitious as a Hyperstore.

THE INFRASTRUCTURE STACK COLLAPSED ENOUGH TO BUILD UPWARDS

Hyperstores rely on the fact that many foundational layers are now standardized:

- container orchestration with Kubernetes
- service meshes
- distributed file systems
- columnar formats
- object storage APIs
- encrypted transport everywhere
- ubiquitous programmable clouds
- compute that scales as fast as storage

The stability of these building blocks made it realistic to create something above them, something integrated, modern, and coherent.

Before this, vendors had to reinvent the lower layers.

Now they can innovate at the architectural level.

THE STACK TAX BECAME UNBEARABLE

Fragmentation was acceptable when:

- AI didn't require context
- privacy regulation was minimal
- applications were simpler
- global multi-tenancy was rare
- data volumes were smaller
- architectures were centralized
- compliance wasn't existential
- developers were cheap relative to infrastructure

None of these conditions remain true.

By the early 2020s, the cost of fragmentation ballooned:

- glue code
- ETL pipelines
- sync processes

- migration risk
- audit failures
- compliance exposure
- divergent schemas
- inconsistent types
- misaligned indexes
- multi-system debugging

Organizations started seeking not "a better database" but "a simpler architecture."

Hyperstores emerged as the answer.

THE MOMENT OF CONVERGENCE HAS ARRIVED

A Hyperstore requires simultaneous maturity across:

- Schema ecosystems
- Cloud infrastructure
- Distributed theory
- Developer tooling
- Cryptography
- AI/ML systems
- Privacy governance frameworks
- Multi-tenancy expectations
- Global-scale storage
- Evented architectures
- Multi-model data patterns

This convergence happened only within the last five years.

Hyperstores were not merely technically infeasible before, **they were conceptually premature.**

The world had to evolve before the architecture could.

CONCLUSION

Hyperstores are possible now because the world finally caught up with the vision.

The layers of technology, culture, regulation, cryptography, and AI have aligned to make vertical data unification both necessary and achievable.

The Hyperstore is not a reaction to database limitations.

It is a response to a global shift in how software itself must operate.

And now that the prerequisites exist, Hyperstores are not just feasible, they are inevitable.

In the next chapter, we formalize the Hyperstore architecture itself, defining its layers, principles, and emergent behaviors.

CHAPTER 7
THE HYPERSTORE ARCHITECTURE

LAYERS, PRINCIPLES, AND BEHAVIORS

The Hyperstore is not a "stack" in the traditional sense. It is a **vertically unified architecture** where storage, indexing, governance, automation, validation, provenance, cryptography, and API generation are all expressions of the same schema-driven system.

Hyperstores replace the fragmented backend with a coherent, layered model, each layer fully aware of and integrated with the others.

This chapter introduces the technical anatomy of a Hyperstore, the principles that govern it, and the behaviors that emerge from its design.

THE ARCHITECTURAL PREMISE: ONE SCHEMA, ONE ENGINE, MANY MODALITIES

A Hyperstore begins with a single schema describing:

- models
- fields
- types

- relations
- indexes
- internationalization
- validation rules
- privacy constraints
- tenant boundaries
- workflows and automations
- provenance requirements
- cryptographic semantics

From this schema, the Hyperstore materializes:

- storage structures
- indexes
- execution rules
- constraints
- API definitions
- SDKs
- migrations
- automations
- access logic

Everything derives from one source of truth. Nothing diverges. Nothing is interpreted differently between layers.

THE LAYERED ARCHITECTURE OF A HYPERSTORE

At its core, a Hyperstore consists of **six tightly integrated layers**:

(1) The Data Layer

The unified substrate for all data modalities:

- Objects (structured data)
- Documents and nested fields
- Graph relationships
- Vector embeddings
- Files and binary payloads
- Derived and computed fields
- Versioned entities

This is not a collection of subsystems, it is one storage architecture capable of expressing multiple data representations natively.

Key Property: **All modalities are first-class citizens.**

No modality is "bolted on."

(2) The Indexing Layer

The indexing layer corresponds directly to the model definitions, creating a unified index topology for:

- structured fields
- full-text content
- i18n variants
- graph edges
- vector embeddings
- hybrid search (text + metadata + vector)

Crucially, these indexes are **coordinated**, not independent.

A change to an entity updates all relevant indexes atomically.

Key Property: **Search, graph traversal, and vector similarity operate over one consistent truth.**

(3) The Governance Layer

A Hyperstore embeds governance directly into its architecture:

- validation
- invariants
- access control
- tenancy boundaries
- field-level privacy
- provenance
- audit trails
- policy enforcement
- lifecycle rules

Governance is not an afterthought, it is encoded as part of the data model and enforced by the engine.

Key Property: **Correctness is guaranteed at the storage boundary, not outsourced to application code.**

(4) The Execution Layer

The Hyperstore includes a native execution environment for:

- automations
- event-driven workflows
- triggers
- scheduled jobs
- cascade operations
- domain events
- projection updates
- embedding refreshes
- derived value computations

Automations run consistently, deterministically, and with direct access to the entity graph and provenance chain.

Key Property: **Behavior becomes a property of the data, not the application.**

(5) The Cryptography Layer

One of the most defining traits of a Hyperstore is its integrated cryptographic substrate:

- post-quantum key exchange (e.g., Kyber)
- post-quantum signatures (e.g., Dilithium)
- homomorphic encryption (numeric domains)
- differential privacy
- encrypted provenance
- per-field encryption semantics
- tenant-isolated key hierarchies
- signature chains across entity versions

The architecture ensures cryptographic guarantees are enforced consistently across modalities.

Key Property: **Security becomes structural, not procedural.**

(6) The Generation Layer

A Hyperstore is fundamentally a **schema-to-system generator**, producing:

- backend methods and endpoints
- typed client SDKs
- model classes
- validation logic
- permission wrappers
- migrations

- serialization helpers
- search utilities
- vector indexing logic

Where traditional systems require developers to write boilerplate, the Hyperstore produces it deterministically.

Key Property: **The backend is generated, not hand-built.**

CROSS-LAYER PRINCIPLES

Several core principles tie the layers together and define what makes a Hyperstore unique:

Multi-Modality Without Fragmentation

Hyperstores reject the premise that different modalities require separate systems.

Instead, they unify them through a shared representation.

Objects, vectors, graphs, files, and provenance metadata all live in the same conceptual space. They share:

- the same entity lifecycle
- the same access policies
- the same validation rules
- the same tenant scope
- the same audit trail
- the same schema

This produces radically simpler applications and dramatically more coherent data models.

Governed by Schema, Enforced by Architecture

In a Hyperstore:

- validation
- permissions
- workflows
- indexes
- relationships
- privacy rules
- automation triggers
- provenance behavior

are *declared*, not *coded*.

The engine enforces correctness.

The application expresses intent.

This eliminates whole classes of errors:

- validation bypasses
- sync failures
- security omissions
- inconsistent policy enforcement
- schema drift
- implicit behavior spread across microservices

The schema becomes the source of architectural truth.

Unified Temporal, Spatial, and Semantic Context

Hyperstores recognize that context is not optional:

- temporal context: versions, lineage

- spatial context: relationships, edges, identity
- semantic context: embeddings, i18n, metadata

Traditional architectures scatter these across multiple stores.

Hyperstores unify them.

The result:

A single entity contains **all its own context**.

Deterministic Construction and Behavior

Everything a Hyperstore produces is deterministic:

- the generated API
- the validation rules
- the indexing behavior
- the access model
- the workflows and triggers
- the provenance records

This eliminates ambiguity and makes systems auditable and predictable, a requirement for regulated sectors and AI-centric workloads.

Data as the Source of Action

In legacy systems, behavior is imposed on data:

- services listen to changes
- scripts run in the background
- orchestrators coordinate flows
- data triggers actions indirectly

Hyperstores invert this model: **Data drives behavior directly.**

Workflows are intrinsic to the data itself. This unlocks coherent domain architectures with no glue code.

EMERGENT BEHAVIORS UNIQUE TO HYPERSTORE ARCHITECTURE

Because of its layered integration, a Hyperstore exhibits emergent behaviors not possible in conventional systems:

Cross-Model Atomicity

A single update may require:

- updating the object
- updating graph edges
- recalculating a vector
- reindexing text
- updating derived fields
- writing provenance
- firing automations

In traditional systems, these operations span multiple databases.

In a Hyperstore, they occur atomically as part of one update cycle.

Zero-Drift Schemas

It becomes impossible for:

- client types to diverge
- API endpoints to contradict the schema
- search indexes to omit fields
- embeddings to misalign
- graph edges to become stale

- provenance and version history to disagree
- migrations to "forget" a constraint

The schema *defines* the system; divergence cannot occur.

Tenant-Isolated Everything

Tenant isolation applies to:

- storage
- indexes
- vectors
- objects
- files
- workflows
- provenance
- cryptographic keys
- access control

No layer can break the boundary.

This makes Hyperstores ideal for SaaS platforms, regulated industries, and multi-tenant AI systems.

AI-Native Data Structures

Because vectors and objects coexist in the same model:

- embeddings update automatically
- semantic search works across all types
- AI automations access complete context
- graph traversals and vector search combine naturally
- text, metadata, and vector signals unify

Hyperstores support AI out-of-the-box without pipelines.

WHY THIS ARCHITECTURE IS A BREAK FROM THE PAST

The Hyperstore architecture is fundamentally different from:

- SQL databases
- document databases
- search engines
- vector databases
- graph stores
- workflow engines
- object storage
- schema registries
- access control systems

Rather than bolting these systems together, a Hyperstore dissolves the boundaries between them.

This is the end of the data stack as a collection of parts.

This is the emergence of data architecture as a single cohesive whole.

In the next chapter, we examine how security, privacy, and trust are enforced structurally within a Hyperstore, rather than delegated to surrounding systems.

CHAPTER 8
THE HYPERSTORE SECURITY MODEL

In conventional data systems, security is layered on top of storage. Encryption is added after schemas are designed. Access control is enforced at the API boundary. Auditability is delegated to external systems. This architecture reflects a historical assumption: that data systems are fundamentally passive, and that trust must be managed elsewhere.

A Hyperstore rejects this assumption.

In a Hyperstore, security is not an add-on, a middleware concern, or a deployment configuration. It is a *structural property* of the system, defined at the same level as data layout and execution semantics. Security guarantees are expressed declaratively, enforced uniformly, and preserved across storage, computation, and access.

This chapter defines the **Hyperstore security model**: its principles, cryptographic foundations, and the guarantees it provides.

THE FAILURE OF PERIMETER SECURITY

Most modern architectures rely on perimeter-based security:

- TLS protects data in transit
- Firewalls protect networks
- IAM systems gate APIs
- Databases assume trusted callers

This model fails in three ways:

1. **Implicit trust** — once inside the perimeter, systems trust each other excessively
2. **Semantic blindness** — encryption protects bytes, not meaning
3. **Audit fragmentation** — provenance, access, and transformation are recorded inconsistently, if at all

As systems become distributed, multi-tenant, and AI-driven, these weaknesses compound. Data is copied, embedded, transformed, and inferred upon — often without cryptographic or semantic continuity.

A Hyperstore addresses this by making security *data-native*.

DECLARATIVE SECURITY GUARANTEES

The defining feature of the Hyperstore security model is that **security guarantees are declared, not inferred**.

Rather than relying on developers to remember where encryption or validation is required, a Hyperstore encodes security requirements directly into its schema and execution rules.

These guarantees apply:

- regardless of access path
- regardless of client language
- regardless of deployment topology

Once declared, they cannot be bypassed without violating the system's invariants.

This approach transforms security from an operational concern into a *mathematical property* of the system.

FIELD-LEVEL ENCRYPTION

Hyperstores operate on structured data, not opaque blobs. This enables **field-level encryption**, where cryptographic protection is applied precisely where sensitivity exists.

Field-level encryption provides several advantages:

- Sensitive fields can be protected without encrypting entire records
- Non-sensitive fields remain queryable and indexable
- Access policies can be defined with semantic precision

Crucially, encryption is associated with *meaning*, not storage location. A field remains encrypted regardless of whether it is accessed via queries, APIs, exports, or derived computations.

HOMOMORPHIC COMPUTATION

A Hyperstore must support computation on protected data. This requirement leads naturally to **homomorphic encryption**.

Using schemes such as Paillier, a Hyperstore can perform specific classes of computation directly on encrypted values without revealing their plaintext. Aggregations, comparisons, and statistical operations can be executed while preserving confidentiality.

This capability enables:

- privacy-preserving analytics
- secure multi-tenant aggregation
- regulatory-compliant data sharing

Homomorphic computation is not treated as an exotic feature. It is an integral part of the Hyperstore's execution model.

DIFFERENTIAL PRIVACY

Beyond encryption, Hyperstores must address inference attacks, situations where sensitive information can be deduced indirectly.

Differential privacy provides a formal framework for controlling such leakage by introducing calibrated noise into query results. In a Hyperstore, differential privacy is applied declaratively, ensuring that:

- privacy budgets are enforced consistently
- repeated queries cannot accumulate excessive information
- guarantees are independent of attacker strategy

This shifts privacy from a best-effort practice to a provable system property.

POST-QUANTUM CRYPTOGRAPHY

Hyperstores are designed for longevity. Data stored today must remain secure against future adversaries.

This requirement motivates native support for **post-quantum cryptography**:

- Post-quantum key exchange (e.g., Kyber) protects data in transit
- Post-quantum digital signatures (e.g., Dilithium) protect integrity and authenticity

By integrating post-quantum primitives at the system level, a Hyperstore avoids the brittle retrofitting that legacy systems will face as cryptographic standards evolve.

PROVENANCE AND CRYPTOGRAPHIC INTEGRITY

Security is not only about secrecy. It is also about **trustworthiness**.

A Hyperstore maintains cryptographic provenance for data:

- hashes ensure integrity
- signatures bind data to identities
- immutability policies prevent unauthorized modification

Because provenance is integrated into the storage engine, it applies uniformly across all operations. Every transformation is traceable. Every state transition is verifiable.

This enables strong audit guarantees without external tooling.

SELECTIVE DISCLOSURE AND ZERO-KNOWLEDGE EXTENSIONS

In many contexts, data must be shared partially. Hyperstores support **selective disclosure**, allowing specific attributes to be revealed while others remain hidden.

Extending this idea further, zero-knowledge techniques enable proofs about data without revealing the data itself. This supports use cases such as:

- identity verification
- compliance attestations
- policy enforcement without disclosure

These techniques reinforce the principle that data systems should reveal *only what is necessary*.

SECURITY ACROSS THE EXECUTION ENGINE

The Hyperstore security model extends through the execution engine:

- Queries respect encryption and access constraints
- Indexes preserve confidentiality guarantees
- Caches do not leak sensitive information
- Write-Ahead Logs maintain integrity and auditability

Security is therefore *end-to-end*, not confined to storage or APIs.

MULTI-TENANCY AND ISOLATION

Multi-tenancy introduces unique security challenges. Hyperstores address these by enforcing isolation at the data, execution, and cryptographic levels.

Tenant boundaries are not advisory; they are structural. Cross-tenant access is impossible without explicit, verifiable authorization.

This makes Hyperstores suitable for environments where multiple organizations, workloads, or trust domains coexist.

SECURITY AS A CATEGORY-LEVEL GUARANTEE

The Hyperstore security model is not defined by individual features, but by its guarantees:

- Confidentiality is enforced at the data level
- Integrity is cryptographically verifiable
- Privacy is provable
- Access is declarative and auditable
- Longevity is built in

These guarantees distinguish Hyperstores from databases, data lakes, and polyglot systems.

SUMMARY

A Hyperstore treats security as a foundational property of data, not a concern of infrastructure or applications.

By integrating encryption, privacy, provenance, and post-quantum resilience into the core architecture, the Hyperstore establishes a new baseline: one in which trust is engineered, not assumed.

This security model is not optional. It is essential to the Hyperstore hypothesis itself.

In the next chapter, we turn to execution, showing how unified planning and deterministic behavior emerge from a single authoritative data engine.

CHAPTER 9
THE HYPERSTORE EXECUTION ENGINE

Traditional databases are fundamentally *storage engines*. Query execution is layered on top as an interpreter over stored structures. In contrast, a Hyperstore is designed around an **execution engine**: a system whose primary responsibility is to evaluate intent over data, regardless of modality, representation, or access path.

This distinction is critical. A Hyperstore does not merely store records and retrieve them efficiently; it executes *unified computational plans* that may involve structured queries, graph traversals, vector similarity search, cryptographic enforcement, provenance validation, and tenant isolation — all within a single execution context.

The Hyperstore execution engine is therefore the locus where the system's guarantees become operational.

UNIFIED QUERY GRAPHS

At the heart of the execution engine is the concept of a **unified query graph**.

Rather than treating relational queries, graph traversals, and vector similarity searches as separate subsystems, a Hyperstore represents all queries as nodes and edges in a single execution graph. Each node represents a logical operation: filtering, joining, traversing, embedding comparison, aggregation; while edges represent data flow and dependency.

This unified representation allows the engine to:

- optimize across modalities
- reorder operations globally
- share intermediate results
- enforce constraints consistently

The result is a system that avoids the impedance mismatches endemic to polyglot architectures.

DETERMINISTIC PLANNING AND EXECUTION

Hyperstores emphasize **determinism** as a core property.

Given the same schema, data state, and query, the execution engine produces the same logical plan. This determinism simplifies reasoning, enables reproducible performance analysis, and strengthens security guarantees.

Deterministic planning is particularly important in the presence of:

- provenance hashing
- cryptographic signatures
- audit requirements
- distributed execution

Non-deterministic execution would undermine these guarantees. The Hyperstore execution engine therefore treats determinism not as an optimization, but as an invariant.

QUERY PLANNING ACROSS MODALITIES

The execution engine's planner operates across all supported data modalities:

- structured data (records and fields)
- object and binary data
- graph relationships
- vector embeddings

Because these modalities are defined within a single schema, the planner has full semantic visibility. It can, for example:

- push filters before vector comparisons
- prune graph traversals using tenant boundaries
- combine scalar predicates with similarity thresholds

This holistic planning capability is not achievable when modalities are handled by separate engines stitched together at the application layer.

INDEX-AWARE EXECUTION

Indexes in a Hyperstore are not passive accelerators. They are active participants in query planning.

The execution engine understands the semantics of:

- hash indexes for exact lookup
- sorted indexes for range queries
- composite indexes for multi-field predicates
- pattern indexes for prefix and wildcard matching
- vector indexes inferred from embedding fields

Because indexes are declared in the schema (or inferred from field properties), the planner can reason about their correctness, cost, and applicability. Index selection is therefore a semantic decision, not a heuristic guess.

WRITE-AHEAD LOGGING AND STATE TRANSITIONS

The Hyperstore execution engine is built on a **Write-Ahead Log (WAL)** that records all state transitions before they are applied.

This log serves multiple purposes:

- durability and crash recovery
- provenance and auditability
- replication and distribution
- deterministic replay

Unlike traditional databases, where the WAL is an implementation detail, the Hyperstore treats the log as a first-class component of execution semantics. Every mutation corresponds to a well-defined state transition in the system.

TRANSACTIONS AND ACID SEMANTICS

Hyperstores provide transactional guarantees appropriate to their unified execution model.

Atomicity, consistency, isolation, and durability are enforced across all modalities involved in a transaction. A transaction that touches structured fields, graph edges, vector indexes, and encrypted attributes is treated as a single logical unit.

This eliminates an entire class of errors common in polyglot systems, where partial failures leave different subsystems in inconsistent states.

CACHING AND MATERIALIZATION

The execution engine employs caching and materialization strategies informed by schema semantics and query structure.

Because queries are represented as graphs, intermediate results can be cached at meaningful boundaries. Cached results respect:

- tenant isolation
- access control
- encryption semantics
- provenance requirements

This ensures that performance optimizations do not compromise correctness or security.

DISTRIBUTED EXECUTION CONSIDERATIONS

Hyperstores are designed to operate in distributed environments.

The execution engine supports:

- partitioned data storage
- distributed index maintenance
- coordinated transaction processing
- deterministic replay across nodes

Crucially, distribution is transparent to the schema and to clients. The same declarative schema governs execution whether the system runs on a single node or across a cluster.

WHY HYPERSTORES ARE MORE OPTIMIZABLE

A recurring theme in this chapter is *semantic visibility*. Because a Hyperstore's schema encodes structure, constraints, and behavior explicitly, the execution engine has more information than a traditional database planner.

This enables optimizations that are impossible in systems where semantics are scattered across application code, ORMs, and external services.

Hyperstores are therefore not merely more capable than polyglot systems; they are inherently more optimizable.

EXECUTION AS A CATEGORY-LEVEL PROPERTY

The Hyperstore execution engine is not an implementation detail of a specific product. It is a defining characteristic of the category.

Any system that claims to be a Hyperstore must provide:

- unified execution across modalities
- deterministic planning
- schema-informed optimization
- transactional integrity across data types

Without these properties, the Hyperstore hypothesis does not hold.

SUMMARY

The Hyperstore execution engine transforms declarative schemas into operational systems.

By unifying query planning, indexing, transactions, and security into a single execution model, it enables a level of correctness, performance, and evolvability that fragmented data architectures cannot achieve.

In a Hyperstore, execution is not an afterthought. It is the realization of the system's theory.

In the next chapter, we describe how these guarantees surface to developers through Fluency, the Hyperstore's schema-derived API contract.

CHAPTER 10
FLUENCY: THE HYPERSTORE API CONTRACT

Most data systems expose *interfaces*: collections of endpoints, query languages, or SDKs that applications must adapt to. These interfaces are often loosely specified, inconsistently implemented, and subject to drift as systems evolve.

A Hyperstore instead exposes a **contract**.

Fluency is the property that this contract is:

- complete
- deterministic
- portable
- language-agnostic
- mechanically derivable from the system definition

In a Hyperstore, clients do not integrate with an API surface. They participate in a shared semantic agreement defined by the schema and enforced by the execution engine.

THE HYPERSTORE AS A PROGRAMMABLE ENDPOINT

A Hyperstore does not present itself as a collection of bespoke services. It presents itself as a *programmable data endpoint* whose behavior is fully described by its schema.

Every externally observable capability — queries, mutations, subscriptions, file transfer, vector search, graph traversal — is governed by the same underlying contract. There is no separate "admin API," "data API," or "search API."

This unification ensures that:

- all access paths observe the same rules
- behavior is predictable
- new capabilities do not fracture the interface

SCHEMA-DERIVED CONTRACTS

The Hyperstore API contract is derived mechanically from the schema. This derivation is not an interpretation or best-effort mapping; it is a deterministic transformation.

The schema defines:

- available models and types
- field-level constraints
- required vs optional data
- access control rules
- provenance and integrity requirements
- encryption and privacy semantics

From this, the Hyperstore generates a complete interaction surface. There is no manual API design step, and therefore *no opportunity for divergence between declared intent and exposed behavior*.

TYPE SYSTEMS AS GUARANTEES

Fluency depends critically on a strong type system.

In a Hyperstore, types are not hints for developers; they are enforceable guarantees. Because all fields are required by default and optionality is explicit, the generated contract has precise semantics:

- clients know exactly which values must be supplied
- nullability is unambiguous
- collections have defined cardinality and uniqueness

This allows generated SDKs to be *correct by construction*, rather than defensively coded against runtime surprises.

DETERMINISTIC CLIENT GENERATION

A core consequence of a schema-derived contract is **deterministic client generation**.

Given the same schema, the Hyperstore will always generate the same logical client surface for a given language and platform. Method names, types, error conditions, and serialization formats are all derived from the schema and execution semantics.

This determinism enables:

- reproducible builds
- cross-team consistency
- confidence in upgrades

Client code becomes a projection of the system, not a handwritten integration.

BINARY PROTOCOLS AND EFFICIENCY

Hyperstores are designed for efficiency as well as expressiveness. The API contract is not bound to text-based protocols alone.

Binary protocols allow:

- compact representation of structured data
- efficient transmission of vectors and binary payloads
- predictable parsing and validation

Because the contract is schema-derived, both client and server share a precise understanding of message structure. This eliminates entire classes of compatibility and performance issues.

ERRORS AS PART OF THE CONTRACT

In many systems, errors are incidental: loosely typed messages returned when something goes wrong.

In a Hyperstore, errors are **part of the contract**.

Validation rules, access control failures, constraint violations, and provenance mismatches all have deterministic representations derived from the schema. Error conditions are therefore:

- enumerable
- localizable
- testable

Clients do not guess what might fail; they are generated with full knowledge of how failure is represented.

STREAMING, SUBSCRIPTIONS, AND REACTIVITY

Fluency extends beyond request–response interactions.

Hyperstores support reactive access patterns: subscriptions to data changes, streaming query results, and event-driven updates; governed by the same contract as synchronous operations.

Because these mechanisms are schema-defined, they respect:

- tenant isolation
- access control
- provenance semantics
- encryption constraints

Reactivity does not introduce a parallel system; it is a natural extension of the contract.

VERSIONING AND EVOLUTION

A contract must evolve without breaking its consumers.

In a Hyperstore, evolution is controlled at the schema level. Changes to types, fields, or constraints are explicit and analyzable. Because client surfaces are generated, compatibility can be assessed mechanically.

This enables:

- safe schema evolution
- controlled deprecation
- parallel version support

Fluency is preserved even as the system grows.

FLUENCY AS A CATEGORY-LEVEL PROPERTY

Fluency is not an implementation detail of a particular product. It is a defining property of the Hyperstore category.

Any system claiming to be a Hyperstore must be able to:

- expose a complete, schema-derived contract
- generate deterministic clients
- enforce semantic guarantees across all access paths
- evolve without interface drift

Without fluency, the promise of a unified data system collapses at the boundary where systems meet users.

SUMMARY

Fluency is how a Hyperstore communicates its guarantees to the outside world.

By elevating schemas into executable contracts and generating deterministic APIs and clients, the Hyperstore eliminates an entire layer of accidental complexity.

The result is not merely a better developer experience, but a system whose correctness extends from storage to execution to integration — coherently, provably, and at scale.

In the next part, we move from theory to instantiation, introducing Vektagraf as a concrete implementation of the Hyperstore model.

PART III

THE IMPLEMENTATION THAT PROVES THE THEORY WORKS

The previous chapters established the Hyperstore as a new category of data system: one defined not by a single access pattern or workload, but by unified semantics, declarative guarantees, and converged execution. What remained, intentionally, was a question of realization.

A category is only real when it can be instantiated.

Vektagraf is that instantiation. It is the world's first system built explicitly to satisfy the Hyperstore definition — not retrofitted from an existing database, not assembled from polyglot components, but designed end-to-end around the Hyperstore hypothesis.

This chapter introduces Vektagraf not as a product feature list, but as a *reference realization* of the Hyperstore model.

If a claim cannot fail, it is not a claim.

CHAPTER 11
VEKTAGRAF: THE FIRST HYPERSTORE
FROM THEORY TO REALITY

No existing database could be extended into a Hyperstore without fundamental compromise.

Relational databases are constrained by table-centric semantics. Document stores fragment behavior into application code. Vector databases optimize similarity search at the expense of transactional guarantees. Graph databases isolate traversal semantics from storage and governance. Even so-called "multi-model" systems remain collections of loosely coupled engines with incompatible invariants.

Vektagraf was created because the Hyperstore requires:

- a unified execution engine
- a single declarative control plane
- first-class support for vectors, graphs, objects, and files
- built-in governance, security, and provenance
- deterministic code generation and fluency

These properties cannot be bolted on after the fact.

THE DESIGN GOALS OF VEKTAGRAF

When Hypermodern set out to build Vektagraf, we started by designing around a small set of non-negotiable principles:

1. **One system, not many** — no sidecars, no external indexes, no split-brain semantics
2. **Schema as architecture** — the schema defines behavior, not just structure
3. **Determinism everywhere** — from execution to code generation
4. **Security by construction** — encryption, privacy, and provenance are native
5. **Fluency by default** — every system is fully accessible through generated, type-safe contracts

These principles map directly to the Hyperstore definition.

CORE CAPABILITIES, UNIFIED

Vektagraf unifies multiple traditionally separate capabilities into a single system:

- structured data storage
- object and binary storage
- vector embeddings and similarity search
- graph relationships and traversal
- full-text and pattern indexing
- multi-tenant isolation
- cryptographic provenance
- declarative validation and automation

What distinguishes Vektagraf is not that it supports each capability individually, but that **they share a single semantic foundation**.

THE VEKTAGRAF SCHEMA AS A HYPERSTORE PROGRAM

At the center of Vektagraf is its schema format.

As described in Part II, a Hyperstore schema is a declarative system definition. Vektagraf implements this idea fully.

Its schemas:

- define types, models, and relationships
- declare validation, automation, and lifecycle rules
- encode security, encryption, and provenance
- specify multi-tenancy and isolation
- drive indexing and performance behavior
- generate APIs, SDKs, tests, and documentation

The schema is compiled, not interpreted. From it, Vektagraf produces a complete, operational system.

EXECUTION: MAKING GUARANTEES REAL

Vektagraf's execution engine realizes the guarantees defined by the Hyperstore model.

Queries involving scalar predicates, graph traversal, vector similarity, and encrypted fields are planned and executed within a unified engine. Transactions span all modalities. Provenance and audit rules are enforced at every state transition.

This execution model is what allows Vektagraf to offer *stronger guarantees* than systems that delegate behavior to application code.

SECURITY AND PRIVACY IN PRACTICE

Vektagraf operationalizes the Hyperstore security model:

- field-level encryption
- homomorphic computation via Paillier
- differential privacy with enforced budgets
- post-quantum key exchange (Kyber)
- post-quantum signatures (Dilithium)

These are not optional plugins. They are native capabilities, integrated into storage, execution, and fluency layers.

MULTI-TENANCY AS AN ARCHITECTURAL PRIMITIVE

In Vektagraf, multi-tenancy is not a convention. It is a structural property.

Tenant isolation is enforced at the schema, execution, and cryptographic layers. Data belonging to different tenants cannot be mixed accidentally, queried implicitly, or leaked through side channels.

This makes Vektagraf suitable for SaaS platforms, regulated environments, and shared infrastructure without compromising isolation guarantees.

FLUENCY REALIZED

Vektagraf implements fluency exactly as defined in Chapter 10.

From a single schema, it generates:

- deterministic APIs
- type-safe client SDKs

- binary and streaming protocols
- validation and error contracts

Clients interact with Vektagraf through a contract, not an ad hoc interface. Integration becomes a mechanical process, not a bespoke engineering effort.

VEKTAGRAF AS A REFERENCE IMPLEMENTATION

Vektagraf should be understood as more than a product. It is a **reference implementation** of the Hyperstore category.

Just as relational theory preceded relational databases, the Hyperstore theory precedes and informs Vektagraf. The system exists to demonstrate that the category is coherent, implementable, and superior to fragmented alternatives.

SUMMARY

Vektagraf makes the Hyperstore real.

It translates theory into architecture, architecture into execution, and execution into fluent interfaces. By doing so, it establishes a new baseline for what a data system can and should be.

With Vektagraf, the Hyperstore moves from hypothesis to fact.

In the next chapter, we examine Vektagraf's internal architecture, showing how Hyperstore principles are realized without fragmentation.

CHAPTER 12
INSIDE VEKTAGRAF
ARCHITECTURE OF A HYPERSTORE

Part II defined the Hyperstore as a category. Chapter 11 introduced Vektagraf as its first concrete realization. This chapter goes one level deeper.

Here, we examine **how Vektagraf is built**, and how its internal architecture maps directly to the theoretical guarantees of a Hyperstore. The goal is not to enumerate components, but to show *why the architecture looks the way it does* — and why alternative designs could not satisfy the same constraints.

Vektagraf's architecture is not accidental. It is the minimum structure required to make the Hyperstore real.

LAYERED, BUT NOT FRAGMENTED

Vektagraf is composed of layers, but it is not a layered system in the traditional sense.

In many databases, layers correspond to separable subsystems: storage here, indexing there, query execution somewhere else. Each layer has its own assumptions, invariants, and failure modes.

In Vektagraf, layers exist for **clarity and responsibility**, not isolation. Each layer is aware of the others through shared semantics defined by the schema.

At a high level, the core layers are:

- the storage layer
- the indexing layer
- the log and WAL layer
- the execution engine
- the schema compiler

These layers cooperate under a single semantic model.

THE UNIFIED STORAGE LAYER

At the foundation of Vektagraf is a **unified storage layer**.

Rather than maintaining separate stores for tables, objects, vectors, and graphs, Vektagraf stores all data as structured, typed entities governed by the same schema rules. Binary data, embeddings, and relational attributes coexist within a single storage abstraction.

This unification enables:

- consistent durability guarantees
- shared provenance and audit trails
- transactional updates across modalities
- elimination of cross-store synchronization logic

Storage is therefore not specialized by data *kind*, but by data *semantics*.

INDEXING AS A SEMANTIC EXTENSION

Indexes in Vektagraf are not auxiliary data structures bolted onto storage. They are **semantic extensions** of the schema.

Scalar indexes (hash, sorted, composite, pattern) are declared explicitly. Vector indexes are inferred directly from vector field definitions. Graph relationships are indexed implicitly through relationship semantics.

Because indexes are schema-driven, the system can guarantee:

- correctness of index usage
- alignment with access control and tenancy rules
- consistency with provenance and encryption semantics

Index maintenance is integrated into the write path, not treated as an asynchronous afterthought.

THE WRITE-AHEAD LOG (WAL)

All state changes in Vektagraf pass through a **Write-Ahead Log**.

The WAL records intent before mutation, ensuring durability, recoverability, and deterministic replay. Importantly, the log is not merely a crash-recovery mechanism. It is a foundational structure that supports:

- transaction atomicity
- provenance verification
- audit reconstruction
- replication and distribution

Every log entry corresponds to a schema-defined operation, preserving semantic meaning even at the lowest level.

THE EXECUTION ENGINE

Above storage and indexing sits the **execution engine** described in Chapter 9.

In Vektagraf, execution is unified across:

- record-level operations
- graph traversals
- vector similarity searches
- encrypted computations
- tenant-isolated queries

The execution engine consumes schema metadata directly, allowing it to plan and optimize queries with full semantic awareness. There is no impedance mismatch between what the schema declares and what the engine executes.

SCHEMA COMPILATION AND SYSTEM GENERATION

One of Vektagraf's defining architectural features is that schemas are **compiled**, not interpreted.

The schema compiler performs:

- structural validation
- dependency resolution across extensions
- expansion of templates and inheritance
- derivation of indexes and constraints
- generation of execution plans and contracts

The output is not just metadata, but a fully realized system configuration that drives storage, execution, security, and fluency.

This compilation step is what enables determinism across environments.

MULTI-TENANCY IN THE ARCHITECTURE

Multi-tenancy permeates the Vektagraf architecture.

Tenant identifiers are treated as first-class attributes at every layer:

- storage layouts respect tenant boundaries
- indexes are tenant-aware
- execution plans enforce isolation
- cryptographic material is tenant-scoped

Because tenancy is architectural rather than conventional, cross-tenant access is impossible without explicit, schema-defined authorization.

CRYPTOGRAPHY AS INFRASTRUCTURE

Encryption, signatures, and privacy controls are embedded directly into the architecture.

Cryptographic operations are not delegated to libraries at the application edge. They are invoked as part of storage writes, query execution, and provenance tracking.

This ensures that cryptographic guarantees hold regardless of how data is accessed or transformed.

WHY THIS ARCHITECTURE IS NECESSARY

It is tempting to ask whether Vektagraf's architecture could have been achieved by composing existing systems.

The answer is no.

The Hyperstore requires:

- unified semantics
- deterministic execution
- schema-driven behavior
- end-to-end guarantees

Any architecture that fragments storage, indexing, execution, or governance into separate systems necessarily fragments guarantees as well.

Vektagraf's architecture exists because the Hyperstore demands it.

SUMMARY

Vektagraf's architecture is the physical realization of the Hyperstore theory.

By unifying storage, indexing, execution, cryptography, and fluency under a single schema-driven model, it achieves guarantees that polyglot and layered systems cannot.

This architecture is not merely innovative. It is inevitable, once the problem is framed correctly.

In the next chapter, we focus on the schema system itself, where structure, validation, governance, and generation converge.

CHAPTER 13
THE VEKTAGRAF SCHEMA SYSTEM

In Vektagraf, the schema is not a description *of* the system. It **is** the system.

Where traditional databases treat schemas as optional structure layered over storage, Vektagraf treats the schema as executable intent. It defines what data exists, how it behaves, how it is secured, how it evolves, and how it is exposed to the world.

This chapter explores the Vektagraf schema system as the concrete realization of the Hyperstore schema model introduced in Part II.

SCHEMA KINDS: EXTENSIONS AND APPLICATIONS

Vektagraf supports two kinds of schemas:

- **Extension schemas**
- **Application schemas**

An *extension schema* defines reusable domain capabilities: models, types, enums, automations, validations, and policies that can be shared across systems.

An *application schema* defines a deployable system. It may declare its own models and types, and it may **import** one or more extension schemas.

This distinction enables composition without fragmentation. Shared logic lives in extensions; concrete systems are assembled at the application layer.

IMPORTS AND COMPOSITION

Application schemas compose behavior by importing extensions.

Imports are explicit and declarative. When an application imports an extension, it incorporates the extension's definitions into its compiled schema graph. Conflicts, overrides, and version mismatches are detected at compile time.

Composition in Vektagraf is therefore:

- deterministic
- analyzable
- safe

This avoids the ambiguity and runtime coupling common in plugin-based systems.

TYPES: PRIMITIVE AND DECLARED

Vektagraf provides a fixed set of **primitive types** — such as id, string, int32, datetime, and vector — which are consumed directly without declaration.

In addition, schemas may declare their own **types**. These declared types encapsulate domain-specific behavior backed by code. From the schema's perspective, there is no distinction in usage between primitive and declared types: both are simply *types*.

Fields may reference:

- a primitive type
- a declared type
- a model
- an enum

The type system is closed, explicit, and fully known at compile time.

MODELS AS SEMANTIC OBJECTS

Models in Vektagraf represent semantic objects, not tables.

A model defines:

- identity
- fields and their types
- required vs optional values
- collection semantics (arrays vs sets)
- constraints and validations
- indexing behavior
- lifecycle and automation hooks

Every field is required unless explicitly marked optional. Collections may be ordered arrays or uniqueness-enforcing sets.

Models are the primary units of storage, execution, and access control.

RELATIONSHIPS WITHOUT FOREIGN KEYS

Vektagraf does not use foreign keys in the traditional sense.

Relationships are expressed by typing fields as other models (or collections of models). Relationship semantics — cardinality, traversal, indexing — are derived from these type declarations.

Because relationships are schema-defined and not inferred from convention, graph traversal and relational access share the same underlying representation.

INHERITANCE AND EXTENSION

Models may extend other models using the extends keyword.

Inheritance allows common structure and behavior to be defined once and reused safely. Extended models inherit fields, constraints, validations, and policies, and may add or refine their own.

This supports polymorphism without sacrificing determinism or clarity.

TEMPLATES AND FIELD INCLUSION

Schemas support reusable field templates through the includes mechanism.

Templates allow a model to include predefined sets of fields, constraints, or behaviors without inheritance. Inclusion is expanded at compile time, ensuring that the final model definition is explicit and complete.

This enables composition without deep inheritance hierarchies.

MULTI-TENANCY AND PROVENANCE

Multi-tenancy and provenance are configured at the schema level.

Top-level settings define whether these features are enabled and how they behave. Individual models may refine provenance behavior, but cannot escape global constraints.

This ensures that governance concerns are addressed uniformly and cannot be bypassed accidentally.

INDEXING AS DECLARATION

Indexes in Vektagraf are declared, not inferred heuristically.

Schemas define index names, fields, types, uniqueness, ordering, and options. Vector indexes are implied by vector field properties such as dimensions and distance functions.

Index definitions are compiled into execution plans and maintained transactionally.

VALIDATION, AUTOMATION, AND LIFECYCLE

Schemas may declare:

- validation rules
- lifecycle hooks
- automated behaviors
- error messages
- test cases

These are executed as part of the system, not as application-side conventions. Data correctness is therefore enforced universally.

INTERNATIONALIZATION AND PRESENTATION

The schema may include labels, descriptions, and localization metadata.

This allows generated clients, documentation, and user interfaces to present data consistently across languages and regions without duplicating logic.

THE SCHEMA COMPILER

All schemas are processed by the Vektagraf schema compiler.

Compilation produces:

- a resolved type graph
- expanded inheritance and templates
- derived constraints and indexes
- execution metadata
- API and SDK contracts

Errors are detected early, before any system is deployed.

WHY THIS MATTERS

By elevating schemas to executable definitions, Vektagraf eliminates the traditional gap between design and implementation.

The schema is no longer documentation. It is the source of truth from which systems are built, verified, and evolved.

SUMMARY

The Vektagraf schema system is the heart of the Hyperstore.

It unifies structure, behavior, governance, and fluency into a single, declarative artifact. In doing so, it transforms schemas from passive descriptions into active, enforceable programs.

In the next chapter, we explore how domain logic, validation, and automation are executed safely and deterministically within the data layer.

CHAPTER 14
AUTOMATIONS, VALIDATION, AND DOMAIN LOGIC

Traditional databases stop at storage and retrieval. Any notion of domain behavior — validation, workflows, invariants, side effects — is pushed into application code.

Vektagraf rejects this split.

In a Hyperstore, **domain logic is declarative, centralized, and enforceable**. Automations, validations, and lifecycle rules are not optional conveniences; they are first-class schema constructs that execute as part of the system itself.

This chapter explains how Vektagraf embeds domain logic directly into the data layer without sacrificing determinism, testability, or safety.

VALIDATIONS AS INVARIANTS

Validations in Vektagraf express *what must always be true* about data.

They may apply at multiple levels:

- field-level constraints (ranges, formats, patterns)

- model-level invariants (cross-field consistency)
- collection constraints (cardinality, uniqueness)
- tenant-scoped rules

Validations are enforced:

- on writes
- on updates
- during automations
- during imports and migrations

Because validations are schema-defined, they are universally applied. No client can bypass them.

DETERMINISTIC VALIDATION EXECUTION

Validation execution is deterministic and ordered.

Given the same schema and input, the same validations run in the same order and produce the same results. This property is critical for:

- reproducible behavior
- consistent error handling
- provenance and audit trails

Validation failures are structured outcomes, not ad hoc exceptions.

ERROR SEMANTICS AND MESSAGES

Errors are part of the schema contract.

Each validation may declare:

- error identifiers

- localized messages
- structured metadata

This allows generated clients to surface meaningful, predictable errors without embedding domain knowledge in application code.

AUTOMATIONS AS DECLARATIVE BEHAVIOR

Automations define *what happens when something changes.*

They may be triggered by:

- create, update, or delete events
- state transitions
- validation outcomes
- scheduled conditions

Automations are declared in the schema and executed by the Hyperstore execution engine. They are not scripts bolted onto the side of the system.

SAFE AND BOUNDED EXECUTION

Automations execute within strict bounds.

They:

- operate on schema-defined data only
- respect tenant isolation
- respect access control
- participate in transactions

This ensures that automations cannot introduce non-deterministic behavior or violate system invariants.

LIFECYCLE HOOKS AND STATE MACHINES

Vektagraf supports lifecycle-aware models.

Schemas may define:

- explicit state fields
- allowed transitions
- transition guards
- side effects tied to transitions

This enables domain state machines to be enforced centrally, rather than reconstructed inconsistently across services.

DOMAIN LOGIC WITHOUT APPLICATION COUPLING

By moving domain logic into the schema:

- business rules are defined once
- all clients observe the same behavior
- migrations preserve invariants
- generated SDKs remain thin and predictable

Application code becomes orchestration, not enforcement.

TESTING AND VERIFICATION

Schemas may include test cases that exercise validations and automations.

These tests are evaluated against the compiled schema and execution engine, ensuring that declared behavior matches actual system behavior.

This enables domain logic to be versioned, tested, and reviewed alongside schema changes.

WHY THIS BELONGS IN THE HYPERSTORE

Automations and validations cannot be reliably enforced at the application edge.

In distributed systems with multiple clients, services, and integrations, only the data system sees *all* writes and *all* state transitions.

The Hyperstore is therefore the only correct place to enforce domain invariants.

SUMMARY

Vektagraf elevates automations, validation, and domain logic to first-class schema concerns.

By embedding behavior directly into the Hyperstore, it eliminates duplication, drift, and inconsistency, and replaces them with deterministic, enforceable rules that scale with the system.

In Vektagraf, data does not merely exist. It behaves, correctly, predictably, and by design.

In the next chapter, we examine privacy not as policy, but as a structural property enforced by the Hyperstore itself.

CHAPTER 15
PRIVACY BY DESIGN

In most systems, privacy is treated as a policy concern layered on top of infrastructure. Data is collected first; controls are applied later. This inversion is the source of many compliance failures and data breaches.

Vektagraf adopts a different stance: **privacy is an architectural property**.

In a Hyperstore, privacy guarantees are encoded directly into schemas, execution semantics, and cryptographic primitives. They are not optional features or operational practices — they are structural constraints that shape what the system can and cannot do.

PRIVACY BY DESIGN VS PRIVACY BY POLICY

Privacy by policy relies on documentation, process, and discipline. Privacy by design relies on enforcement.

Vektagraf implements privacy by design by ensuring that:

- sensitive data is explicitly identified

- access rules are schema-defined
- transformations are constrained
- leakage paths are eliminated by construction

If a privacy rule cannot be enforced by the Hyperstore, it is considered incomplete.

DATA MINIMIZATION IN THE SCHEMA

The schema is the first line of defense.

By declaring:

- required vs optional fields
- scoped collections
- tenant boundaries
- retention and lifecycle rules

Vektagraf ensures that systems collect only the data they intend to collect. Accidental over-collection becomes structurally difficult.

Data minimization is therefore a compile-time concern, not merely a runtime guideline.

FIELD-LEVEL CONFIDENTIALITY

Privacy in Vektagraf operates at the field level.

Schemas may mark fields as confidential, restricted, or protected by cryptographic controls. These declarations govern:

- who may read or write the field
- whether the field may be indexed
- whether it may participate in queries
- how it is serialized in APIs

Confidentiality is enforced uniformly across all access paths.

ENCRYPTION AS DEFAULT, NOT EXCEPTION

Encryption is not an add-on in Vektagraf.

Sensitive fields may be declared as encrypted, with encryption enforced at rest, in transit, and — where applicable — during computation. Because encryption semantics are schema-defined, they cannot be bypassed by clients or operators.

This eliminates the class of failures where data is "supposed to be encrypted" but is not.

SELECTIVE DISCLOSURE

Not all consumers of data require the same visibility.

Vektagraf supports selective disclosure by allowing schemas to define:

- redacted views
- role-scoped projections
- derived representations

Clients receive only the data they are entitled to see, in the form they are permitted to see it. Raw data does not leak through convenience APIs.

DIFFERENTIAL PRIVACY AND AGGREGATION

For analytical and aggregate queries, Vektagraf supports differential privacy mechanisms.

Schemas may define privacy budgets, noise strategies, and

allowable aggregations. The execution engine enforces these constraints automatically.

This allows insights to be extracted without exposing individual-level data, a capability increasingly required by regulation.

TENANT ISOLATION AS PRIVACY BOUNDARY

Multi-tenancy is not merely an operational convenience; it is a privacy boundary.

In Vektagraf, tenant isolation is enforced at every layer:

- storage
- indexing
- execution
- cryptography

Cross-tenant access is impossible without explicit, schema-defined authorization. Accidental leakage through shared infrastructure is structurally prevented.

PROVENANCE AND ACCOUNTABILITY

Privacy requires accountability.

Vektagraf's provenance system records:

- who accessed data
- what transformations occurred
- when state changed
- under which rules

These records enable auditing, investigation, and compliance verification without relying on external logging systems.

RIGHT TO ERASURE AND LIFECYCLE CONTROL

Privacy regulations increasingly require support for deletion, anonymization, and retention control.

Vektagraf supports lifecycle rules that define:

- retention periods
- erasure conditions
- anonymization strategies

These rules are enforced by the system itself via automation, ensuring that privacy obligations are met consistently and verifiably.

PRIVACY AS A HYPERSTORE PROPERTY

Privacy by design is not unique to Vektagraf. It is a natural consequence of the Hyperstore model.

When schemas are executable, execution is deterministic, and security is native, privacy stops being aspirational and becomes enforceable.

SUMMARY

Vektagraf embeds privacy into the fabric of the system.

By making privacy rules declarative, enforceable, and inseparable from execution, it enables systems that are compliant by construction rather than compliant by exception.

In a Hyperstore, privacy is not something you promise. It is something you prove.

In the next chapter, we show how multi-modal querying becomes possible when objects, graphs, vectors, and text share a single execution context.

CHAPTER 16
MULTI-MODEL QUERYING IN VEKTAGRAF

In conventional architectures, querying is fragmented by modality. SQL is used for structured data, bespoke traversal languages for graphs, vector APIs for similarity search, and separate services for full-text or object retrieval. Each modality introduces its own syntax, semantics, and limitations.

Vektagraf eliminates these silos.

As a Hyperstore, Vektagraf supports **multi-model querying** as a single, unified capability. Queries may span structured fields, relationships, vectors, files, and encrypted attributes within one coherent execution plan.

QUERIES AS UNIFIED INTENT

A Vektagraf query expresses *intent*, not mechanism.

Rather than instructing the system how to traverse indexes or which engine to invoke, queries describe:

- the data of interest
- the constraints that apply

- the relationships to follow
- the similarity or relevance criteria

The execution engine determines how this intent is satisfied, using the schema and indexes to plan an optimal, deterministic execution.

STRUCTURED QUERIES

Structured querying in Vektagraf resembles familiar paradigms but with stricter semantics.

Field predicates, ranges, ordering, grouping, and aggregation are expressed over models and fields defined in the schema. Because all fields are typed and required by default, query semantics are unambiguous.

Structured queries benefit from:

- declared indexes
- deterministic execution plans
- transactional consistency
- tenant-aware filtering

GRAPH TRAVERSAL AS NATIVE QUERYING

Relationships in Vektagraf are first-class schema constructs, not inferred joins.

Queries may traverse relationships directly, expressing depth, direction, and constraints without switching languages or engines. Graph traversal participates in the same execution plan as structured filtering.

Because traversal semantics are schema-defined:

- access control is enforced at each hop

- tenant boundaries are respected
- provenance is preserved

Graph querying is therefore not a special case, but a natural extension of object querying.

VECTOR SIMILARITY SEARCH

Vektagraf supports vector embeddings as native field types.

Queries may include similarity conditions such as nearest-neighbor search, distance thresholds, or relevance ranking. Vector operations can be combined with scalar predicates and graph constraints in a single query.

For example, a query may:

- restrict results to a tenant
- filter by structured attributes
- traverse a relationship
- rank results by vector similarity

All of this is planned and executed as one operation.

FULL-TEXT AND PATTERN MATCHING

Textual search is integrated alongside other modalities.

Schemas may declare text fields with indexing and language options. Queries may include prefix, substring, pattern, or relevance-based matching, combined with other constraints.

Text search does not require exporting data to an external search engine; it is part of the same execution model.

QUERYING ENCRYPTED AND PROTECTED FIELDS

Some fields are encrypted or protected by privacy constraints.

Vektagraf allows limited, schema-approved operations on such fields, including:

- equality checks
- range queries (where supported)
- aggregate computation via homomorphic encryption

Queries that violate declared security or privacy rules are rejected deterministically. There is no implicit downgrade or silent failure.

COMPOSABILITY ACROSS MODALITIES

The defining property of Vektagraf's querying model is **composability**.

Modalities are not layered; they are orthogonal. Any combination of structured filters, traversals, similarity conditions, and aggregations may be expressed — subject to schema constraints.

This composability enables expressive queries that would otherwise require multiple systems and significant application-side orchestration.

DETERMINISTIC PLANNING AND OPTIMIZATION

All multi-model queries are planned deterministically.

The planner considers:

- index availability
- modality-specific costs

- security and tenancy constraints
- data locality and distribution

Because the query graph is unified, optimizations can cross modality boundaries. For example, a scalar filter may reduce the candidate set before vector similarity is evaluated.

TRANSACTIONS AND CONSISTENCY

Multi-model queries observe a consistent view of the system.

Queries executed within a transaction see a snapshot that is coherent across structured data, relationships, vectors, and derived indexes. There is no risk of reading partially updated state from different subsystems.

This consistency is essential for correctness in complex applications.

WHY THIS MATTERS

Multi-model querying is often advertised as a convenience feature.

In Vektagraf, it is a structural capability enabled by the Hyperstore architecture. Because all data shares a schema, execution engine, and security model, expressive queries remain safe, deterministic, and optimizable.

This is not achievable by stitching together specialized databases.

SUMMARY

Vektagraf's multi-model querying allows applications to ask richer questions of their data, without fragmentation, duplication, or loss of guarantees.

By unifying structured, graph, vector, and protected data querying into a single execution model, Vektagraf demonstrates one of the clearest practical advantages of the Hyperstore approach.

In a Hyperstore, querying is no longer constrained by data modality. It is constrained only by intent and correctness.

In the next chapter, we evaluate the performance characteristics of a Hyperstore, and why conventional benchmarks fail to capture its capabilities.

CHAPTER 17
BENCHMARKS AND PERFORMANCE CHARACTERISTICS

Benchmarking a Hyperstore requires a different frame of reference.

Traditional benchmarks assume narrowly defined workloads: transactional throughput, analytical scans, graph traversal depth, or vector similarity latency, each measured in isolation, often on systems optimized for that single task.

Vektagraf does not fit this model.

As the first Hyperstore, Vektagraf is designed to execute **multi-model, governed, transactional workloads** within a single system. Its performance characteristics must therefore be understood in terms of *capability density*, *semantic overhead*, and *end-to-end efficiency*, not raw single-axis throughput alone.

This chapter explains how to evaluate Vektagraf's performance honestly and meaningfully.

WHY CONVENTIONAL BENCHMARKS FALL SHORT

Standard benchmarks such as TPC-C, TPC-H, vector recall tests, or graph traversal microbenchmarks isolate one dimension of behavior.

When applied to Vektagraf, these benchmarks miss critical factors:

- schema-driven validation and enforcement
- tenant isolation
- encryption and privacy constraints
- provenance tracking
- unified execution planning

A naive comparison may show higher per-operation cost, while ignoring that equivalent guarantees would require *multiple systems* and significant application logic elsewhere.

MEASURING CAPABILITY DENSITY

A more appropriate metric for Hyperstores is **capability density**: the amount of validated, governed, and queryable behavior delivered per unit of infrastructure.

Vektagraf consolidates workloads that would otherwise require:

- a transactional database
- a vector database
- a graph database
- an object store
- an authorization service
- an audit and provenance system

Measured this way, overall system performance often improves due to reduced data duplication, fewer network hops, and elimination of synchronization overhead.

BASELINE PERFORMANCE CHARACTERISTICS

At a baseline level, Vektagraf exhibits performance characteristics comparable to modern, well-engineered databases for equivalent operations:

- indexed scalar lookups execute in low-millisecond or sub-millisecond ranges
- range queries and aggregations scale with declared indexes
- vector similarity search performance is dominated by index choice and dimensionality
- graph traversal performance reflects relationship density and depth

These characteristics are achieved while maintaining transactional integrity and governance guarantees.

MULTI-MODEL QUERY PERFORMANCE

The distinguishing performance trait of Vektagraf emerges in **composed queries**.

Queries that combine structured filtering, relationship traversal, and vector similarity are executed within a single plan. This avoids:

- intermediate materialization between systems
- redundant serialization and deserialization
- network latency between services

In practice, this often yields lower end-to-end latency than decomposed pipelines built on specialized databases.

IMPACT OF GOVERNANCE AND SECURITY

Governance features impose real costs, and Vektagraf makes these costs explicit.

Validation, encryption, provenance hashing, and access checks add measurable overhead. However, because these operations are native and schema-driven, their cost is predictable and bounded.

Crucially, equivalent guarantees implemented at the application layer are typically:

- slower due to duplication
- harder to optimize
- inconsistently enforced

Vektagraf trades hidden, unbounded cost for visible, engineered overhead.

INDEXING AND OPTIMIZATION EFFECTS

Because indexes are declared and schema-aware, Vektagraf can optimize aggressively without risking semantic violations.

This enables:

- early pruning of candidate sets
- reduced vector search domains
- efficient traversal planning
- safe reuse of cached intermediates

Over time, systems benefit as schemas mature and indexing strategies are refined.

SCALING CHARACTERISTICS

Vektagraf is designed to scale along multiple axes:

- data volume
- tenant count
- query complexity
- vector dimensionality

Horizontal scaling distributes storage, indexes, and execution while preserving deterministic behavior. Because the schema governs partitioning and isolation, scaling does not weaken guarantees.

INTERPRETING BENCHMARK RESULTS

When evaluating Vektagraf benchmarks, the correct question is not:

> Is this faster than a specialized database?

but rather:

> What would it take to achieve the same guarantees with other systems?

Once governance, security, correctness, and integration costs are accounted for, Vektagraf's performance profile becomes competitive, and often superior, at the system level.

TOWARD STANDARD HYPERSTORE BENCHMARKS

The emergence of Hyperstores suggests the need for new benchmark classes.

Such benchmarks would measure:

- multi-model query latency
- transactional consistency across modalities
- governance enforcement cost
- developer and operational complexity

Vektagraf provides a concrete foundation upon which such benchmarks can be defined.

SUMMARY

Vektagraf's performance cannot be reduced to a single number.

Its strength lies in executing complex, governed, multi-model workloads efficiently within one system, replacing fragile, distributed pipelines with a coherent execution model.

Benchmarked correctly, Vektagraf demonstrates that the Hyperstore is not only more expressive than existing data systems, but performant enough to make that expressiveness practical.

In the next part, we step beyond implementation to examine the first classes of applications that Hyperstores make possible.

PART IV

CASE STUDIES, REAL-WORLD USES, AND FORWARD-LOOKING ANALYSIS

Up to this point, this manifesto has focused on *what a Hyperstore is* and *how Vektagraf realizes it*. Architecture, schemas, execution, security, and performance establish the foundation of a new data category.

This chapter shifts perspective.

Here we explore what becomes *possible* when those guarantees are treated as givens rather than aspirations. These are not hypothetical demos or narrowly scoped examples. They are classes of applications that are impractical, fragile, or prohibitively complex on conventional data stacks, but become natural when built on a Hyperstore.

What systems make easy
determines what they make inevitable.

CHAPTER 18
THE FIRST HYPERSTORE APPLICATIONS

AI-powered search typically requires a pipeline of systems:

- a primary database
- an embedding generation service
- a vector database
- a relevance or ranking layer
- an access-control filter applied post hoc

Each boundary introduces latency, inconsistency, and risk.

In a Hyperstore, AI search is **native**.

With Vektagraf:

- embeddings are stored alongside source data
- vector similarity participates in transactional queries
- structured filters, graph context, and tenancy are applied *before* ranking
- access control and privacy are enforced during execution, not after

The result is search that is faster, safer, and semantically richer, without an external pipeline.

RECOMMENDATION ENGINES WITHOUT DATA PIPELINES

Recommendation systems are traditionally built on batch jobs, feature stores, and offline materialization.

Vektagraf enables real-time recommendation queries that combine:

- user behavior (structured data)
- relationships (graphs)
- content similarity (vectors)
- business rules and constraints (schema logic)

Because all modalities are queryable together, recommendations can be generated *on demand*, using live data, while respecting privacy and tenancy rules.

This eliminates the lag and brittleness of pipeline-driven systems.

HEALTHCARE AND PRIVACY-CRITICAL SYSTEMS

Healthcare systems face a uniquely difficult combination of requirements:

- strict privacy regulation
- fine-grained access control
- auditability and provenance
- secure analytics over sensitive data

Hyperstores are uniquely suited to this domain.

With Vektagraf:

- patient data can be encrypted at the field level
- access is constrained by schema-defined roles
- analytics can be performed using differential privacy or homomorphic techniques
- every access and transformation is provably auditable

This enables data sharing and insight without violating confidentiality, something legacy systems struggle to achieve.

MULTI-TENANT SAAS PLATFORMS

Most SaaS platforms are multi-tenant by necessity, but not by design.

Tenant isolation is often enforced through conventions, query filters, or duplicated infrastructure. These approaches are error-prone and difficult to audit.

In Vektagraf, multi-tenancy is structural.

Applications built on a Hyperstore benefit from:

- schema-enforced tenant boundaries
- tenant-aware indexing and execution
- cryptographically scoped data
- predictable performance as tenant count grows

This allows SaaS platforms to scale securely without multiplying complexity.

KNOWLEDGE GRAPHS WITHOUT A SEPARATE SYSTEM

Knowledge graphs are often implemented as standalone databases, disconnected from operational data.

Vektagraf collapses this distinction.

Because relationships are first-class schema constructs, operational data *is already a graph*. Queries may traverse relationships, apply constraints, and incorporate similarity search without exporting data to a separate graph engine.

This enables living knowledge graphs that evolve transactionally with the system they describe.

ENTERPRISE SECURITY AND GOVERNANCE PLATFORMS

Security platforms depend on correlating diverse signals:

- identities and roles
- events and logs
- relationships between entities
- behavioral similarity and anomaly detection

These workloads are inherently multi-model.

Vektagraf supports them natively, allowing security systems to:

- query across structured events, graphs, and vectors
- enforce least-privilege access by schema
- maintain immutable audit trails
- operate across multiple tenants or organizations

This makes the Hyperstore a natural foundation for next-generation security infrastructure.

WHY THESE APPLICATIONS EMERGE FIRST

These application classes share common traits:

- high semantic complexity

- strong governance requirements
- multi-model data access
- sensitivity to correctness and privacy

They are precisely the workloads that strain traditional data architectures.

The Hyperstore does not merely optimize them, it *simplifies* them by removing entire layers of accidental complexity.

THE PATTERN: FEWER SYSTEMS, STRONGER GUARANTEES

Across all these examples, a consistent pattern emerges.

Hyperstore-based applications:

- replace pipelines with queries
- replace conventions with guarantees
- replace glue code with schema
- replace operational risk with determinism

This is the practical payoff of the Hyperstore model.

SUMMARY

The first Hyperstore applications are not niche experiments. They are foundational systems that demand correctness, expressiveness, and trust.

By unifying data, behavior, security, and execution, Vektagraf makes these applications simpler to build and safer to operate.

What was once an architectural burden becomes a natural consequence of the system.

With this chapter, the Hyperstore moves from theory, to implementation, to impact.

In the next chapter, we formalize what Hyperstores enable that databases, by design, cannot.

CHAPTER 19
WHAT HYPERSTORES ENABLE THAT DATABASES CANNOT

It is tempting to view Hyperstores as simply more capable databases: faster, more flexible, more modern.

This framing is incorrect.

The difference between a database and a Hyperstore is not a matter of incremental improvement. It is a difference of *kind*. Hyperstores enable entire classes of systems and guarantees that databases, by their architecture, cannot provide, regardless of scale, tuning, or extensions.

This chapter articulates those differences explicitly.

SEMANTIC-NATIVE SYSTEMS

Databases store data; meaning lives elsewhere.

In conventional systems, semantics are scattered across:

- application code

- ORMs and query builders
- documentation and tribal knowledge
- operational conventions

The database enforces structure, but not meaning.

A Hyperstore is **semantic-native**.

In Vektagraf, meaning is declared in the schema and enforced by the system itself. Types, relationships, invariants, lifecycle rules, and access constraints are not advisory, they are executable.

This enables systems where:

- correctness is structural
- behavior is predictable
- reasoning does not depend on external code

Databases cannot provide this without relinquishing control to applications.

PRIVACY-ENFORCED COMPUTATION

Databases can encrypt data. They cannot *compute safely on it.*

Privacy-aware computation in conventional systems relies on fragile discipline:

- filtering after retrieval
- masking in application code
- separate analytics environments

Hyperstores enforce privacy during execution.

With schema-defined encryption, selective disclosure, differential privacy, and homomorphic operations, a Hyperstore ensures that:

- forbidden computations cannot be expressed
- permitted computations cannot leak excess information
- guarantees hold regardless of client behavior

This is not achievable in systems where execution semantics are unaware of privacy intent.

CONTINUOUS SCHEMA GOVERNANCE

In most systems, governance is episodic.

Schemas evolve informally. Constraints drift. Policies are enforced unevenly. Audits occur after the fact.

Hyperstores enable **continuous governance**.

Because the schema is compiled, versioned, and executable:

- changes are explicit and reviewable
- invariants remain enforced across versions
- migrations preserve meaning, not just shape
- violations are impossible to ignore

Governance becomes an ongoing property of the system, not an external process.

ZERO-INTEGRATION ARCHITECTURES

Modern systems suffer from integration overload.

Each new capability: search, vectors, graphs, authorization, auditing; introduces another service, API, and failure mode.

Hyperstores collapse these boundaries.

By unifying modalities under a single execution engine and schema, Hyperstores enable **zero-integration architectures**:

- no ETL pipelines between core capabilities
- no synchronization logic
- no duplicated data models
- no cross-system consistency gaps

The system integrates with itself by construction.

UNIFIED CONCURRENCY MODELS

Concurrency in polyglot systems is fractured.

Transactions apply in one database but not another. Eventual consistency is tolerated because coordination is impractical. Developers reason about race conditions across services rather than within a coherent model.

Hyperstores provide a **unified concurrency model**.

Transactions, isolation, and ordering apply across all data modalities and behaviors:

- structured updates
- graph mutations
- vector index changes
- automations and lifecycle hooks

This allows developers to reason about system state holistically, rather than defensively.

ELIMINATION OF ACCIDENTAL COMPLEXITY

Many of the hardest problems in modern systems are not intrinsic to the domain.

They arise from:

- impedance mismatches between tools

- duplicated enforcement logic
- inconsistent guarantees
- partial failure modes

Hyperstores eliminate entire categories of this accidental complexity by making guarantees structural.

What remains is the *essential* complexity of the problem being solved.

WHY DATABASES CANNOT CATCH UP

Could databases evolve to provide these capabilities?

Incrementally, no.

The database model assumes:

- passive storage
- application-owned semantics
- externalized governance
- fragmented execution

Retrofitting semantic execution, privacy-aware computation, and unified modalities would require breaking these assumptions; effectively becoming a Hyperstore in all but name.

At that point, the category has changed.

THE STRATEGIC IMPLICATION

Hyperstores shift where complexity lives.

Instead of distributing correctness across dozens of services and teams, they concentrate it into a single, declarative system.

This has profound implications:

- faster system design
- lower operational risk
- stronger compliance posture
- longer system longevity

These are strategic advantages, not mere technical conveniences.

SUMMARY

Hyperstores enable systems that databases fundamentally cannot.

By being semantic-native, privacy-enforcing, continuously governed, integration-free, and concurrency-unified, they redefine what a data system is responsible for.

Vektagraf demonstrates that these capabilities are not theoretical. They are practical, implementable, and transformative.

The question is no longer whether Hyperstores are possible, but how long systems can afford to remain without them.

In the next chapter, we analyze the economic and engineering consequences of replacing fragmented stacks with unified architectures.

CHAPTER 20
ECONOMIC AND ENGINEERING IMPACT

New data architectures are often evaluated narrowly: latency curves, throughput numbers, or feature matrices. These metrics matter, but they do not capture the *systemic* consequences of architectural choice.

Hyperstores change the economics and engineering dynamics of software systems.

By unifying data modalities, execution semantics, governance, and security into a single system, Vektagraf alters cost structures, team workflows, risk profiles, and long-term maintainability. This chapter examines those impacts directly.

COST COMPRESSION VIA UNIFICATION

Modern data stacks are expensive not because individual components are inefficient, but because **capability is fragmented.**

A typical production system may require:

- a transactional database
- a vector database

- a graph database
- an object store
- a search engine
- an authorization service
- an audit and logging pipeline

Each introduces infrastructure cost, operational overhead, and failure modes.

A Hyperstore collapses these into a single system.

With Vektagraf:

- data is stored once
- indexes are shared
- execution is unified
- governance is native

The result is *cost compression* — fewer machines, fewer services, fewer engineers required to keep the system reliable.

DEVELOPER PRODUCTIVITY AND COGNITIVE LOAD

Engineering productivity is constrained less by language syntax than by **cognitive load.**

In fragmented architectures, developers must reason about:

- multiple data models
- inconsistent guarantees
- partial failures
- synchronization and eventual consistency

Hyperstores reduce this burden.

By making schemas executable and guarantees structural, Vektagraf allows developers to:

- define behavior once
- rely on system enforcement
- reason about data holistically

Productivity increases not because developers work faster, but because they work with fewer conceptual surfaces.

FASTER TIME TO CORRECTNESS

Most engineering effort is spent not on building features, but on fixing inconsistencies.

Race conditions, authorization bugs, data drift, and schema mismatches consume disproportionate time.

Hyperstores shorten the path to correctness.

Because validation, access control, and lifecycle rules are enforced centrally:

- entire classes of bugs never occur
- edge cases are surfaced early
- behavior is consistent across clients

Correctness becomes the default state, not an achievement earned through exhaustive testing.

REDUCED SECURITY BLAST RADIUS

Security failures are often systemic.

When data is copied across systems, a breach in one component can expose far more information than intended. Authorization filters applied late in pipelines magnify risk.

Vektagraf reduces the security blast radius by design:

- data is not duplicated across systems
- access control is enforced at execution
- encryption is field-level and mandatory
- tenant isolation is structural

A compromise in one area does not cascade across the architecture.

OPERATIONAL SIMPLICITY

Operational complexity scales superlinearly with the number of systems.

Each additional service requires:

- deployment pipelines
- monitoring and alerting
- backup and recovery
- on-call expertise

By unifying capabilities, Hyperstores simplify operations.

Teams operate *one system* with consistent semantics, rather than coordinating many with mismatched behaviors. This reduces operational load and improves reliability.

MAINTAINABILITY AND LONGEVITY

Most systems fail not because they cannot scale, but because they cannot *evolve*.

Schemas drift. Assumptions break. Institutional knowledge erodes.

Hyperstores counteract this through:

- explicit, versioned schemas
- deterministic behavior
- centralized domain logic
- analyzable evolution paths

Vektagraf systems remain understandable years after their creation, even as teams and requirements change.

ORGANIZATIONAL IMPACT

Architecture shapes organizations.

Fragmented systems encourage siloed teams, defensive coding, and slow coordination. Unified systems encourage shared understanding and clearer ownership.

Hyperstores shift responsibility from integration glue to domain modeling, a more stable and valuable form of engineering effort.

THE COMPOUNDING EFFECT

The benefits of a Hyperstore compound over time.

Reduced complexity lowers the cost of change. Lower cost of change encourages iteration. Iteration improves product quality and resilience.

This feedback loop is difficult to achieve in architectures burdened by integration debt.

SUMMARY

The impact of Hyperstores extends beyond technical elegance.

By compressing cost, increasing productivity, reducing risk, and extending system lifespan, Vektagraf alters the economics of building and operating complex software.

This is not merely an optimization. It is a structural shift in how value is created, and preserved, in data-driven systems.

In the next chapter, we look forward, examining how Hyperstores reshape system design, organizational practice, and the future of data infrastructure.

CHAPTER 21
THE FUTURE OF HYPERSTORES

Every new systems category follows a familiar arc: skepticism, experimentation, early adoption, and eventual normalization. The Hyperstore is now at the beginning of that trajectory.

Vektagraf demonstrates that the category is not only coherent, but practical. The remaining question is not *whether* Hyperstores will exist, but *how they will reshape the landscape* as they mature.

This chapter looks forward, not to speculate idly, but to outline plausible, technically grounded futures.

NEAR-TERM PREDICTIONS

In the near term, Hyperstores are likely to appear first where architectural pain is most acute:

- AI-heavy systems struggling with fragmented data pipelines
- privacy- and compliance-critical platforms
- multi-tenant SaaS operating at scale

- knowledge-centric systems with evolving schemas

Early adopters will not replace their entire stack at once. Instead, Hyperstores will enter as **consolidation points**, absorbing responsibilities previously handled by multiple services.

THE DECLINE OF THE POLYGLOT DEFAULT

For over a decade, the industry has treated polyglot persistence as inevitable.

Hyperstores challenge that assumption.

As unified systems demonstrate that they can handle diverse workloads *without sacrificing guarantees*, the default architectural question will shift from:

> *Which databases do we need?*

to:

> *What semantics should our data system enforce?*

Polyglot stacks will persist at the edges, but they will no longer be the assumed baseline.

RESEARCH AVENUES

The Hyperstore category opens multiple research directions:

- richer semantic optimization in unified query planners
- advanced privacy-preserving computation techniques
- formal verification of schema-defined invariants
- adaptive indexing across modalities

- long-term cryptographic agility and migration

These are not incremental database improvements. They sit at the intersection of databases, programming languages, cryptography, and systems theory.

TOWARD STANDARDIZATION

As with relational databases, standardization will follow implementation.

Potential areas for convergence include:

- schema definition languages
- semantic query representations
- provenance and audit models
- privacy and encryption declarations
- contract and client generation formats

Standardization will not eliminate competition; it will accelerate it by allowing multiple implementations to share a conceptual foundation.

OTHER HYPERSTORE IMPLEMENTATIONS

Vektagraf is unlikely to remain the only Hyperstore.

As the category becomes understood, other implementations will emerge:

- open-source research systems
- vertically optimized Hyperstores for specific domains
- cloud-native Hyperstores integrated into platforms

These systems may differ in performance characteristics, deployment models, or supported cryptographic primitives, but they will share the defining properties outlined in Part II.

THE SHIFT IN ENGINEERING PRACTICE

When Hyperstores become mainstream, engineering practice will change.

Developers will spend less time:

- wiring systems together
- duplicating validation logic
- compensating for inconsistent guarantees

and more time:

- modeling domains precisely
- reasoning about system behavior
- evolving schemas intentionally

Data modeling will regain its central role as a core engineering discipline.

ORGANIZATIONAL AND EDUCATIONAL IMPACT

As Hyperstores reshape architectures, they will also reshape organizations.

Teams will orient around *domains* rather than *infrastructure components*. Education will emphasize semantic modeling, privacy-aware design, and declarative system thinking.

The boundary between "application" and "data layer" will become less rigid, not because responsibilities blur, but because guarantees are shared.

A WORLD WITH MAINSTREAM HYPERSTORES

In a world where Hyperstores are commonplace:

- data systems are fewer, but more capable
- correctness is assumed, not heroic
- privacy and security are structural, not reactive
- integration is the exception, not the norm

Systems become easier to reason about, safer to evolve, and more resilient over time.

SUMMARY

The Hyperstore represents a shift in how we think about data systems, from passive storage to active semantic infrastructure.

Vektagraf shows what is possible today. The future will refine, expand, and diversify the category.

As Hyperstores move from novelty to norm, they will quietly redefine expectations, until building systems any other way feels unnecessarily fragile.

That is how new categories win: not by shouting, but by making the old ones obsolete.

In the next chapter, we address the most common objections and misconceptions, and test the Hyperstore thesis against rigorous skepticism.

CHAPTER 22
OBJECTIONS, MISCONCEPTIONS, AND REBUTTALS

Every genuinely new systems category encounters resistance. This resistance is rarely irrational; it reflects accumulated experience with overhyped technologies, premature abstractions, and tools that promised unification but delivered complexity.

Hyperstores challenge deeply held assumptions about how data systems should be built. It is therefore reasonable, and necessary, to examine the objections directly.

This chapter addresses the most common misconceptions and critiques, and explains why they do not hold when applied to a true Hyperstore.

"THIS IS JUST ANOTHER MULTI-MODEL DATABASE"

The objection: Hyperstores appear to be multi-model databases with better marketing, supporting documents, graphs, and vectors in one product.

The rebuttal: Multi-model databases aggregate engines. Hyperstores unify semantics.

In a multi-model system:

- each modality has its own execution model
- guarantees differ by access path
- governance is uneven
- cross-model queries are limited or bolted on

In a Hyperstore:

- all modalities share a single execution engine
- transactions span all operations
- security and privacy are enforced uniformly
- queries are planned holistically

The difference is architectural, not cosmetic.

"THIS BELONGS IN APPLICATION CODE"

The objection: Validations, workflows, access rules, and semantics should live in application logic, not the data system.

The rebuttal: Application code does not see all writes.

In distributed systems with multiple services, clients, and integrations, only the data system observes every state transition. Any invariant enforced outside it is, by definition, partial.

Hyperstores move *enforcement* to the only layer that can guarantee it, while keeping *orchestration* in applications.

This separation increases correctness without reducing flexibility.

"SCHEMAS WILL BECOME TOO COMPLEX"

The objection: Encoding behavior, security, and governance in schemas risks creating monolithic, unreadable definitions.

The rebuttal: Complexity already exists; schemas make it explicit.

In conventional systems, the same complexity is spread across:

- database schemas
- ORMs
- services
- validation libraries
- authorization frameworks

Hyperstore schemas consolidate this into a single, versioned artifact. Modular schemas, imports, inheritance, and templates exist precisely to manage scale.

Explicit complexity is preferable to hidden complexity.

"PERFORMANCE WILL SUFFER"

The objection: Unifying so many responsibilities into one system must introduce unacceptable overhead.

The rebuttal: Hyperstores trade *visible, bounded cost* for *hidden, unbounded cost.*

Yes, enforcing validation, encryption, and governance adds overhead. But equivalent guarantees implemented across multiple systems incur:

- repeated serialization
- network latency
- synchronization delays

- defensive checks at every boundary

In practice, end-to-end performance often improves once pipelines collapse into single execution plans.

"THIS IS TOO OPINIONATED"

The objection: Hyperstores impose a particular way of modeling data and behavior, limiting freedom.

The rebuttal: All systems are opinionated. The difference is whether opinions are implicit or explicit.

Hyperstores are opinionated about *guarantees*, not about domains. They enforce correctness, determinism, and governance, but leave domain modeling entirely to the schema author.

Freedom without guarantees is not flexibility; it is deferred failure.

"IT WON'T FIT LEGACY SYSTEMS"

The objection: Existing systems are too entrenched to adopt a Hyperstore.

The rebuttal: Hyperstores are not an all-or-nothing proposition.

They integrate at boundaries through generated contracts and APIs. Early adoption often targets:

- new capabilities
- greenfield services
- consolidation of fragmented subsystems

Over time, as confidence grows, responsibilities migrate

inward. This is how new infrastructure categories always enter mature environments.

"THIS IS REINVENTING THE DATABASE"

The objection: Hyperstores seem to discard decades of database research.

The rebuttal: Hyperstores build on that research, they do not reject it.

Indexes, transactions, logs, query planning, and storage engines all remain. What changes is *where meaning lives* and *what guarantees are enforced by the system rather than assumed by applications.*

Relational theory did not replace filesystems; it subsumed them. Hyperstores play a similar role.

"ONLY LARGE TEAMS WILL BENEFIT"

The objection: The sophistication of Hyperstores only pays off at large scale.

The rebuttal: Smaller teams benefit disproportionately.

When resources are limited, eliminating entire classes of bugs, integrations, and security risks matters more, not less. Hyperstores allow small teams to build systems with guarantees previously reserved for large organizations.

"THIS SOUNDS TOO GOOD TO BE TRUE"

The objection: Unified systems have failed before; this will too.

The rebuttal: Past failures conflated *unification of features* with *unification of semantics.*

Hyperstores succeed because they start with theory: schemas as executable intent, deterministic execution, and security as structure. Vektagraf demonstrates that these ideas can be implemented coherently.

Skepticism is healthy. Dismissing new categories because old ones failed is not.

SUMMARY

Hyperstores provoke objections because they challenge comfortable assumptions.

When examined closely, these objections resolve into concerns about complexity, performance, and control; concerns that Hyperstores address structurally rather than dismissing.

Vektagraf does not ask users to suspend disbelief. It invites scrutiny.

And under scrutiny, the Hyperstore model holds.

In the epilogue, we return to first principles, situating the Hyperstore within the broader arc of the Hypermodern Theorem.

EPILOGUE

THE HYPERSTORE THESIS

This book began with a constraint and ended with an architecture.

The constraint was simple: **systems cannot be reliable if truth is reconciled after execution**. The architecture that follows: the Hyperstore, exists to eliminate that failure mode by making data unified, authoritative, and executable by construction.

This is why this book is a manifesto.

It does not propose optimizations.

It does not recommend best practices.

It redraws the boundary of responsibility.

For decades, the industry accepted that meaning lived in applications, that governance lived in policy, that trust lived in process, and that data systems were merely storage engines beneath it all.

That separation no longer holds.

WHAT HAS BEEN ESTABLISHED

If the arguments of this book are correct, then several conclusions follow mechanically:

- Fragmented data architectures cannot support deterministic AI.
- Privacy-by-policy is structurally weaker than privacy-by-design.
- Multi-tenancy enforced in application code is inherently fragile.
- Provenance recorded externally cannot be made authoritative.
- Developer discipline is not a substitute for architectural guarantees.

The Hyperstore exists to absorb those responsibilities into the data layer itself, where they can be enforced, audited, and reasoned about.

Vektagraf demonstrates that this is not theoretical.

WHY THIS IS ONLY THE BEGINNING

This book addresses **data** because data is the base layer.

But the Hypermodern Theorem does not stop there.

Once data becomes unified and authoritative, two additional domains become tractable in ways they never were before:

- **Authority** — how systems decide who may act, decide, or assert truth.
- **Coordination** — how systems align actions across agents, organizations, and time without central fragility.

Those domains cannot be solved on fragmented data. They require a substrate where state is coherent, provenance is intrinsic, and execution is deterministic.

That substrate is established here.

The remaining books do not speculate upward. They build on this foundation.

A MANIFESTO IS NOT A CONCLUSION

Manifestos are not written to end arguments.

They are written to make certain arguments unavoidable.

If the Hyperstore is unnecessary, this book should be refuted architecturally.

If unified data cannot be authoritative without governance, it should be demonstrated concretely.

If fragmentation is still sufficient, it should be proven under modern constraints.

If none of those refutations hold, then the industry is already transitioning, whether it has named the category or not.

This book gives that transition a name.

It defines the base layer.

Everything that follows depends on it.

ABOUT THE AUTHOR

Dom Jocubeit is a software architect and founder-engineer working at the intersection of data systems, security, and applied AI.

He is the originator of the *Hyperstore* category, a unified data system model that treats schemas as executable intent and enforces semantics, privacy, and governance at the system level. He is also the creator of *Vektagraf*, the first Hyperstore implementation, which serves as a concrete, falsifiable test of the category's feasibility.

Through this work, Dom developed the **Hypermodern Theorem**, a set of falsifiable claims about how large-scale systems must be constructed once human-scale assumptions no longer hold. The theorem identifies data, authority, and coordination as domains that must be resolved mechanically rather than socially for systems to remain coherent under automation, anonymity, and scale. Each claim is instantiated through a working system designed to fail publicly if the underlying thesis does not hold.

Dom is the founder and Chief Engineer of Hypermodern AI, where he designs and builds production-grade systems that collapse fragmented data stacks into coherent, semantically enforced platforms. His work spans distributed systems, cryptography-aware architectures, and large-scale application design, with an emphasis on determinism and execution over governance and interpretation.

Prior to Hypermodern, Dom developed full-stack and infrastructure systems across both startup and enterprise environments, with deep technical roots in distributed system design. His focus has consistently been on correctness, longevity, and architectural clarity over short-term optimization.

The **Hyperstore Manifesto** is the first articulation of the Hypermodern Theorem. It synthesizes Dom's technical work into a formal boundary, introduces a new systems category, and establishes the conditions required for deterministic execution at scale.

Dom's work is driven by a single conviction: *that software systems should become simpler by being stricter, and that when meaning and guarantees are built into the foundation, both technology and the people who build it can move faster with confidence.*

linkedin.com/in/jocubeit
x.com/jocubeit

GLOSSARY

HYPERSTORE TERMS

ACID
Atomicity, Consistency, Isolation, Durability. In a Hyperstore, ACID guarantees apply uniformly across all modalities (objects, graphs, vectors, automations), not just tabular data.

Automation
A schema-defined, system-executed behavior triggered by data lifecycle events (create, update, delete, transition). Automations are deterministic and transactional.

Blast Radius
The scope of impact when a failure, breach, or bug occurs. Hyperstores reduce blast radius by centralizing enforcement and minimizing data duplication.

Compiled Schema
A schema that is parsed, validated, versioned, and transformed into executable plans enforced by the Hyperstore at runtime.

Continuous Governance
The property of enforcing policies, constraints, and invariants at all times through the schema, rather than via periodic audits or external processes.

Deterministic Execution
Execution where the same inputs, schema, and state always produce the same outputs and side effects. Essential for auditability and reasoning.

Executable Intent
The idea that schemas express not only structure but meaning, rules, and guarantees that the system enforces directly.

Field-Level Encryption
Encryption applied at individual field granularity, declared in the schema and enforced by the Hyperstore.

Graph Engine
The component responsible for relationship traversal, connectivity queries, and graph-aware execution within a unified engine.

Hyperstore
A unified data system that enforces semantic, security, and execution guarantees across multiple data modalities through an executable schema.

Hyperstore Hypothesis
The claim that unifying semantics, execution, and governance into a single system enables capabilities that databases fundamentally cannot provide.

Imports
A schema mechanism for composing and reusing definitions across extensions, enabling modularity and versioned evolution.

Invariant
A rule that must always hold true for system state. In Hyperstores, invariants are enforced centrally and cannot be bypassed.

Lifecycle Rule
A schema-defined constraint governing how data may change over time, including state transitions and retention policies.

Multi-Model Query
A single query plan that spans structured data, graph traversal, and vector similarity under unified semantics.

Polyglot Persistence
An architectural pattern using multiple specialized databases. Hyperstores aim to eliminate the need for this by unification.

Primitive Type
A built-in type provided by the Hyperstore runtime (e.g., id, string, int32, datetime) that does not require schema declaration.

Privacy-Enforced Computation
Computation where privacy constraints are enforced during execution, preventing unauthorized inference or leakage by design.

Provenance
Metadata describing the origin, transformation, and history of data, maintained automatically by the Hyperstore.

Schema as Ontology
The principle that the schema defines the authoritative meaning of data within the system, not merely its shape.

Semantic-Native
A system property where meaning is first-class and enforced, rather than inferred or reconstructed in application code.

Selective Disclosure
The ability to reveal only authorized subsets or derivations of data, enforced structurally at execution time.

Tenant Isolation
Structural separation of data and execution between tenants, guaranteed by the Hyperstore rather than by convention.

Transaction Boundary
The scope within which operations are atomic and isolated. In Hyperstores, transaction boundaries span all modalities and automations.

Unified Execution Engine
A single engine responsible for planning and executing all operations: queries, mutations, automations, and index updates.

Vector Engine
The component responsible for similarity search and embedding-aware operations, integrated into the unified execution model.

Zero-Integration Architecture
An architectural outcome where capabilities do not require external synchronization, pipelines, or glue code because they are unified by design.